计算机系列教材

钟 静 熊 江 主 编
冯宗玺 闫东方 吴 愚 副主编

计算机专业英语

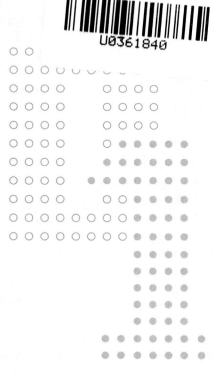

清华大学出版社
北京

内 容 简 介

本书共分9个单元,涉及计算机科学基础知识、操作系统和计算机体系结构、软件工程、信息管理、网络应用、物联网、人工智能、信息安全等内容,除了传统的计算机基本原理介绍,也包含当前比较热门的技术,如云计算、信息存储、3D互联网、人工智能、元宇宙等,各单元还介绍了有关计算机专业英语的学习方法、词汇构成和翻译技巧等知识。全书以计算机和IT领域的最新英语时文和经典原版教材为基础,配以详尽的注释和练习,使读者能够快速掌握计算机专业英语的一般特点和大量专业词汇。书后附有词汇表和缩略词表。

本书适合作为高等院校计算机及相关专业本科生计算机专业英语的教材,同时可供IT行业相关工程人员以及其他需要学习计算机专业英语的读者参考。

本书封面贴有清华大学出版社防伪标签,无标签者不得销售。
版权所有,侵权必究。举报: 010-62782989,beiqinquan@tup.tsinghua.edu.cn。

图书在版编目(CIP)数据

计算机专业英语/钟静,熊江主编. —北京:清华大学出版社,2022.8(2025.2 重印)
计算机系列教材
ISBN 978-7-302-61413-5

Ⅰ.①计… Ⅱ.①钟… ②熊… Ⅲ.①电子计算机-英语-教材 Ⅳ.①TP3

中国版本图书馆 CIP 数据核字(2022)第 134017 号

责任编辑:张　玥
封面设计:常雪影
责任校对:李建庄
责任印制:刘海龙

出版发行:清华大学出版社
网　　址: https://www.tup.com.cn,https://www.wqxuetang.com
地　　址:北京清华大学学研大厦A座　　邮　编:100084
社 总 机:010-83470000 邮　购:010-62786544
投稿与读者服务:010-62776969,c-service@tup.tsinghua.edu.cn
质量反馈:010-62772015,zhiliang@tup.tsinghua.edu.cn
课件下载: https://www.tup.com.cn,010-83470236

印 装 者:三河市龙大印装有限公司
经　　销:全国新华书店
开　　本:185mm×260mm　印　张:11.75　字　数:273千字
版　　次:2022年9月第1版　印　次:2025年2月第4次印刷
定　　价:49.80元

产品编号:097199-01

前　言

　　学习最新的计算机技术，操作计算机软硬件产品，进行软件开发或网络应用，都离不开对计算机专业英语的熟练掌握。尤其是近几年，随着云计算、大数据、人工智能、互联网技术的不断发展，越来越多的新兴技术不断涌现，影响着我们的生活和工作。我国在高科技领域也逐步成熟起来，不断涌现出具有影响力的自主产品。我们在作者编写的《计算机英语》教材的基础上编写了这本教材，增加了超级计算机、量子计算机、互联网＋、5G、人工智能、鸿蒙操作系统、元宇宙等内容。

　　本书共分 9 个单元，包括计算机概览、操作系统和计算机体系结构、软件工程、信息管理、计算机网络、因特网、物联网、人工智能和信息安全。每个单元分 3 部分。第一部分是精读课文，可以在教师的辅导下在课堂上完成；第二部分是泛读课文，学生可以自行阅读，以扩充该单元涉及的知识；第三部分也是泛读部分，但其中穿插了一些计算机专业英语学习指南，如如何学习计算机专业英语、计算机专业英语的词汇构成和计算机专业英语的翻译技巧，帮助学生轻松、全面地掌握计算机专业英语。

　　本教材有以下主要特色：

　　（1）精选难度适中的课文。教材中的阅读材料参考了经典原版教材和 IT 领域的最新英语时文。所有课文既反映计算机科学的技术概貌，又紧扣技术潮流。课文内容简明易读、知识实用、图文并茂。

　　（2）提供详尽的注释、注音和学习指南。教材分两栏排版，右栏是课文，加粗显示了该掌握的或需要学习的单词和专业术语；左栏详尽地标注出相应课文中生词的音标和中文释义。所有的音标和中文释义均参考 2010 年 12 月由外语教学与研究出版社出版的《牛津·外研社英汉汉英词典》，音标按照英式发音标注。对于课文中出现的缩略词以及有关地点、重要人物都在页下加了注释。学习指南中介绍了如何学习计算机专业英语、计算机专业英语词汇构成法和计算机专业英语的翻译技巧，以更好地帮助学生学习、掌握好计算机专业英语。

　　（3）选材广泛。为了让学生对计算机技术有较全面的认识，本教材既有关于计算机发展的历史介绍、硬件构成等基础内容，也有云计算、物联网、人工智能、元宇宙等新技术，既有硬件产品手册，也有软件开发指南。其目的是使学生全面地掌握计算机领域的专业英语。

　　（4）激励爱国情怀。教材选取了我国自主研发的产品、技术，如华为 5G、"神威·太湖之光"超级计算机等，也有中国 IT 公司介绍，如华为、联想、中科曙光、浪潮等。培养学生的家国情怀、时代责任以及创新精神。

　　本教材的教学时数为 30 学时左右。

　　本教材的第 1 单元由熊江编写，第 2～5 单元由钟静编写，第 6 单元由冯宗玺编写，第 7 单元由闫东方编写，第 8 单元由吴愚编写，第 9 单元由吴鸿娟编写。全书由钟静统稿。

尽管作者在资料查核、术语翻译等方面做了大量工作，但由于计算机领域的发展日新月异，许多新术语尚无确定的规范译法，加上时间仓促，作者水平有限，书中难免有不妥之处，恳请广大读者及时提出宝贵意见和建议。

作　者

2022 年 3 月

目　　录

Unit One　Computer Overview ··· 1
　　Section A　About Computer ··· 1
　　Section B　Storage ·· 9
　　Section C　如何学习计算机专业英语 ··· 14

Unit Two　Operating System and Computer Architecture ············· 18
　　Section A　Operating System ··· 18
　　Section B　Computer Architecture ·· 29
　　Section C　计算机专业英语词汇 ·· 33

Unit Three　Software Engineering ·· 38
　　Section A　Software Engineering Methodologies ······················· 38
　　Section B　Exploratory Testing ··· 44
　　Section C　计算机专业英语翻译 ·· 47

Unit Four　Information Management ··· 51
　　Section A　Information Storage ·· 51
　　Section B　Data Mining ·· 59
　　Section C　Data Center ·· 63

Unit Five　Networking ··· 67
　　Section A　Networking ··· 67
　　Section B　Distributed System ·· 74
　　Section C　Software Configuration Guide For Cisco 2600
　　　　　　　 Series Routers ·· 79

Unit Six　Internet ··· 86
　　Section A　Internet ·· 86
　　Section B　5G ··· 93
　　Section C　Top 10 Search Engines in the World ······················· 99

Unit Seven   The Internet of Things 104
    Section A   Cloud Infrastructure and Services 104
    Section B   The Internet of Things 115
    Section C   Supercomputer 122

Unit Eight   Artificial Intelligence 127
    Section A   Artificial Intelligence 127
    Section B   Metaverse 136
    Section C   Robotics 143

Unit Nine   Computer Security 149
    Section A   Network Security 149
    Section B   Digital Signature 155
    Section C   Smartphone Security 159

词汇表 163

缩略词表 176

参考文献 180

Unit One Computer Overview

Section A About Computer

Computer science is the **discipline** that seeks to build a scientific foundation for such topics as computer design, computer programming, information processing, **algorithmic** solutions of problems, and the algorithmic process itself. It provides the underpinnings for today's computer applications as well as the foundations for tomorrow's computing **infrastructure**.

I. History

- **Abacus**

Today's computers have an extensive **genealogy**. One of the earlier computing devices was the abacus. It most likely had its roots in ancient China and was used in the early Greek and Roman civilizations. The machine is quite simple, consisting of beads strung on rods that are in turn mounted in a rectangular frame. As the beads are moved back and forth on the rods, their positions represent stored values. It is in the positions of the beads that this "computer" represents and stores data. For control of an algorithm's execution, the machine relies on the human **operator**. Thus the abacus alone is merely a data storage system.

discipline
/ˈdɪsɪplɪn/
n. 学科；纪律
algorithmic
/ˌælɡəˈrɪðmɪk/
adj. 算法的
infrastructure
/ˈɪnfrəˌstrʌktʃə/
n. 基础设施

genealogy
/ˌdʒiːnɪˈælədʒɪ/
n. 系统学；系谱
operator
/ˈɒpəreɪtə/
n. 操作员

- **Adding Machine**

After the Middle Ages and before the Modern Era the quest for more sophisticated computing machines was seeded. A few inventors began to experiment with the technology of **gears**. Among these were Blaise Pascal① of France, Gottfried Wilhelm Leibniz② of Germany, and Charles Babbage③ of England. The first adding machine, a precursor of the **digital** computer, was devised in 1642 by Blaise Pascal. This device employed a series of ten-toothed wheels, each tooth representing a digit from 0 to 9, with data being input **mechanically** by establishing initial gear positions. Output was achieved by observing the final gear positions.

- **Analytical Engine**

In the 19th century, the British mathematician and inventor Charles Babbage conceived the **Difference** Engine (of which only a **demonstration** model was constructed) could be modified to perform a variety of calculations, but his **Analytical** Engine was designed to read instructions in the form of holes in paper cards. Thus Babbage's Analytical Engine was programmable. In fact, Augusta Ada Byron④, who published a paper in which she demonstrated how Babbage's Analytical Engine could be programmed to perform various computations, is often identified today as the world's first programmer.

- **Electronic Computers**

The technology of the time was unable to produce the complex gear-driven machines of Pascal, Leibniz, and Babbage in a financially feasible manner. But with the advances in **electronics** in the early 1900s, this **barrier** was overcome. Some researchers were applying the technology of **vacuum** tubes to construct totally

---

① Blaise Pascal:布莱士·帕斯卡(1623—1662),法国数学家、物理学家、思想家。
② Gottfried Wilhelm Leibniz:戈特弗里德·威廉·莱布尼茨(1646—1716),德国哲学家、逻辑学家、数学家和科学家,数理逻辑的先驱,提出了二进制。
③ Charles Babbage:查尔斯·巴贝奇(1792—1871),英国数学家和发明家,可编程计算机的发明者,计算机先驱。
④ Augusta Ada Byron:奥古斯塔·埃达·拜伦(1815—1852),英国数学家,诗人拜伦的女儿。她建立了循环和子程序概念,被视为世界上第一位软件设计师。

---

gear
/ɡɪə/
n. 齿轮
digital
/ˈdɪdʒɪtl/
adj. 数字的
mechanically
/mɪˈkænɪkəlɪ/
adv. 机械地

difference
/ˈdɪfərəns/
n. 差数;差别
demonstration
/ˌdemənˈstreɪʃən/
n. 演示;论证
analytical
/ˌænəˈlɪtɪkəl/
adj. 分析的

electronics
/ˌɪlekˈtrɒnɪks/
n. 电子学;电子器件
barrier
/ˈbærɪə/
n. 障碍
vacuum
/ˈvækjuəm/
n. 真空

electronic computers. The first of these machines was apparently the Atanasoff-Berry machine①, constructed during the period from 1937 to 1941 at Iowa State College by John Atanasoff and his assistant, Clifford Berry②. Another was a machine called **Colossus**, used by the team headed by Alan Turing③ to **decode** German messages during the latter part of World War II. Other, more **flexible** machines, such as the **ENIAC**④(electronic numerical **integrator** and **calculator**) developed(see Figure 1A-1).

Figure 1A-1　ENIAC

- **Integrated Circuits**

The history of computing machines has been closely linked to advancing technology, including the invention of **transistors** and the subsequent development of **circuits** constructed as single units, called **integrated** circuits. With these developments, the room-sized machines of the 1940s were reduced over the decades to the size of single **cabinets**, while the processing power of computing machines began to double every two years. As work on integrated circuitry progressed, many of the circuits within a computer became readily available on the open market as integrated circuits **encased** in toy-sized blocks of plastic called **chips**.

---

① Atanasoff-Berry machine: 阿塔纳索夫-贝瑞计算机,简称 ABC。
② constructed during the period from 1937 to 1941 at Iowa State College by John Atanasoff and his assistant, Clifford Berry: 由爱荷华州立大学的约翰·文森特·阿塔纳索夫和他的研究生克利福特·贝瑞在 1937 年至 1941 年间开发。
③ Alan Turing: 阿兰·图灵(1912—1954),英国数学家和逻辑学家,提出了"图灵机"和"图灵测试"等重要概念,被誉为"计算机科学之父"和"人工智能之父"。
④ ENIAC: 电子数字积分计算机。

Ⅱ. Modern Computer

• **Desktop**

A major step toward popularizing computing was the development of desktop computers.

In 1981, IBM① introduced its first desktop computer, called the **personal computer**, or **PC**, whose underlying software was developed by a newly formed company known as Microsoft②. The PC was an **instant** success. Today, the term PC is widely used to refer to all those machines (from various manufacturers) whose design has evolved from IBM's initial desktop computer. At times, however, the term PC is used interchangeably with the generic terms **desktop** or **laptop**.

• **Internet**

As the twentieth century drew to a close, the ability to connect individual computers in a world-wide system called the **Internet** was revolutionizing communication. A British scientist Tim Berners-Lee③ proposed a system by which documents stored on computers throughout the Internet could be linked together producing a **maze** of linked information called the World Wide Web(shortened to "Web"). To make the information on the Web accessible, software systems, called search engines, were developed to "**sift** through" the Web, "**categorize**" their findings, and then use the results to assist users researching particular topics. Major players in this field are Google④, Yahoo⑤, and Microsoft. These companies continue to expand their Web-related activities, often in directions that challenge our traditional way of thinking.

PC
个人计算机
instant
/'ɪnstənt/
n. 瞬间
desktop
/'desktɒp/
n. 台式机
laptop
/'læptɒp/
n. 笔记本电脑

maze
/meɪz/
n. 错综复杂;迷惑
sift
/sɪft/
v. 审查;筛查
categorize
/'kætɪɡəraɪz/
vt. 分类

---

① IBM：International Business Machines Corporation,(美国)国际商用机器公司。
② Microsoft：美国微软公司。有 Microsoft Windows 和 Microsoft Office 系列软件。
③ Tim Berners-Lee：蒂姆·伯纳斯·李爵士(1955年出生于英国)是万维网的发明者，互联网之父,1990年12月25日,他在日内瓦的欧洲粒子物理实验室里开发出了世界上第一个网页浏览器。
④ Google：中文译名"谷歌",创建于1998年9月,被公认为全球最大的搜索引擎。
⑤ Yahoo：中文译名"雅虎",1994年创立于美国。全球第一家提供因特网导航服务的网站。

- **Embedded**

As the desktop computers were being accepted and used in homes, the **miniaturization** of computing machines continued. Today, tiny computers are **embedded** within various devices. For example, automobiles now contain small computers running Global Positioning Systems(GPS), monitoring the function of the engine, and providing voice command services for controlling the car's audio and phone communication systems.

Ⅲ. Smartphone

Perhaps the most potentially revolutionary application of computer miniaturization is found in the expanding capabilities of **portable** telephones. Indeed, what was recently merely a telephone has evolved into a small hand-held general purpose computer known as a smartphone(see Figure 1A-2) on which telephony is only one of many applications. These "phones" are equipped with a rich array of **sensors** and interfaces including cameras, microphones, **compasses**, touch screens, **accelerometers** (to detect the phone's orientation and motion), and a number of wireless technologies to communicate with other smartphones and computers. The potential is enormous. Indeed, many argue that the smartphone will have a greater effect on society than the PC.

Figure 1A-2   Smartphone

The miniaturization of computers and their expanding capabilities have brought computer technology to the forefront of today's society. Computer technology is so prevalent now that **familiarity** with it is fundamental to being a member of modern society. Computing technology has altered the ability of governments to **exert** control; had enormous impact on global economics; led to startling advances in scientific research; revolutionized the role of data collection, storage, and applications; provided new means for people to communicate and interact; and has repeatedly challenged society's status **quo**. The result is a **proliferation** of subjects surrounding computer science, each of which is now a significant field of study in its own right. Indeed, an introduction to computer science is an interdisciplinary undertaking.

Ⅳ. Quantum Computer

In classical computers, the unit of information is called a "bit" and can have a value of either 1 or 0. But its equivalent in a **quantum** system—the **qubit**(quantum bit)—can be both 1 and 0 at the same time. This phenomenon opens the door for multiple calculations to be performed simultaneously. But the qubits need to be **synchronised** using a **quantum effect** known as **entanglement**, which Albert Einstein termed "spooky action at a distance[①]".

However, scientists have struggled to build working devices with enough qubits to make them competitive with conventional types of computer. Google, the technology giant's Sycamore[②] contains 54 qubits, although one of them did not work, so the device ran on 53 qubits. Sycamore quantum processor was able to perform a specific task in 200 seconds that would take the world's best supercomputers 10 000 years to complete.

In the University of Science and Technology of China, Chinese scientists have established the quantum computer **prototype**, named "Jiuzhang"[③] (see Figure 1A-3), via which up to 76 photons were

---

① spooky action at a distance: 爱因斯坦提出的"鬼魅般的超距作用"。
② Sycamore: 谷歌开发的54比特量子芯片名,于2019年发布。
③ Jiuzhang: 九章。2020年12月4日,中国科学技术大学宣布成功构建76个光子的量子计算原型机"九章"。这一突破使我国成为全球第二个(第一个为谷歌的Sycamore)实现"量子优越性"(国外称"量子霸权")的国家。

detected. The photo below released on Dec. 4, 2020 shows a quantum computer prototype operates.

Figure 1A-3   "Jiuzhang" the Quantum Computer Prototype

Jiuzhang's quantum computing system can implement large-scale GBS 100 trillion times faster than the world's fastest existing supercomputer. This achievement marks that China has reached the first milestone on the path to full-scale quantum computing—a quantum computational advantage.

Exercises

Ⅰ. Fill in the blanks with the information given in the text.

1. _____ provides the underpinnings for today's computer applications as well as the foundations for tomorrow's computing infrastructure.

2. The first adding machine, a precursor of the _____ computer, was devised in 1642 by Blaise Pascal.

3. In fact, _____ is often identified today as the world's first programmer.

4. Integrated circuits encased in toy-sized blocks of plastic called _____.

5. In 1981, IBM introduced its first desktop computer, called _____.

6. The ability to connect individual computers in a world-wide system called the _____ was revolutionizing communication.

7. Smartphones are equipped with a rich array of _____ and interfaces including cameras, microphones, compasses, touch screens, accelerometers and a number of wireless technologies.

8. _____ technology has altered the ability of governments to exert control; had enormous impact on global economics.

II. Translate the following terms or phrases from English into Chinese.

algorithm	encode
programming	software
hardware	computing
the personal computer, or PC	Internet
operator	vacuum tube
transistor	integrated circuit
chip	World Wide Web
search engine	embed
smartphone	Global Positioning Systems(GPS)
sensor	quantum bit
quantum effect	

III. Translate the following passage from English into Chinese.

Huawei WATCH GT3

On November 17, 2021, Huawei unveiled its smart WATCH GT3 with a 46mm dial, 1.43 inch AMOLED HD color screen, 11mm thick, weight as low as 42.6g and a native Hongmon Harmony OS operating system.

This digital wristwatch features a rotating crown and a microphone, is synced to a cellphone, allowing users to answer calls and receive text messages from their wrists. It can track workouts and use an array of Apps, support GPS, Beidou, GLONASS, Galileo and QZSS positioning, stable and anti-interference.

Huawei WATCH GT3 series has added altitude blood oxygen monitoring. In the altitude care mode, the WATCH can monitor changes in altitude and blood oxygen saturation, and provide risk assessment of altitude sickness, respiratory training guidance and health advice.

Section B Storage

Computers information is encoded as **patterns** of 0s and 1s. These **digits** are called bits(short for **binary** digits) which are really only symbols whose meaning depends on the application at hand. The patterns of bits are used to represent numeric values, characters in an alphabet and **punctuation** marks, images, or sounds.

For the purpose of storing data, a computer contains a large collection of circuits(such as **flip-flops**), each capable of storing a single bit. This bit **reservoir** is known as the machine's main memory which is organized in manageable units called cells, with a typical cell size being eight bits. And a string of eight bits is called a byte. To identify individual cells in a computer's main memory, each cell is assigned a unique "name," called its address.

Because a computer's main memory is organized as individual, addressable cells, the cells can be accessed independently as required. To reflect the ability to access cells in any order, a computer's main memory is often called random access memory (**RAM**).

Due to the **volatility** and limited size of a computer's main memory, most computers have additional memory devices called mass storage systems, including magnetic disks, CDs, DVDs, magnetic tapes, and **flash drives**. The advantages of mass storage systems over main memory include less volatility, large storage capacities, low cost, and in many cases, the ability to remove the storage medium from the machine for **archival** purposes.

A major disadvantage of mass storage systems is that they typically require mechanical motion and therefore require significantly more time to store and retrieve data than a machine's main memory, where all activities are performed electronically.

- **Magnetic Systems**

For years, magnetic technology has dominated the mass storage arena. Magnetic storage that magnetically records data on the flat surfaces of one or more disks that rotate around a common shaft and has sets of magnetic read-write heads that write and read the recorded data.

- **Optical Systems**

Another class of mass storage systems applies optical technology. An example is the compact disk(CD). These disks are 12 centimeters(approximately 5 inches) in diameter and consist of **reflective** material covered with a clear protective coating. Information is recorded on them by creating **variations** in their reflective surfaces. This information can then be **retrieved** by means of a laser beam that detects **irregularities** on the reflective surface of the CD as it spins.

Traditional CDs have capacities in the range of 600 to 700MB. However, DVDs(Digital Versatile Disks), which are constructed from precisely focused laser, provide storage capacities of several GB. Such disks are capable of storing lengthy multimedia presentations, including entire motion pictures. Finally, **Blu-ray** technology, which uses a laser in the blue- **violet spectrum** of light (instead of red), is able to focus its laser beam with very fine precision. As a result, **BDs**(Blu-ray Disks) provides over five times the capacity of a DVD(see Figure 1B-1). This seemingly vast amount of multiple, semitransparent layers that serve as distinct surfaces when viewed by a storage is needed to meet the demands of high definition video.

- **Flash Drives**

Flash drives are capable of up to a few hundred GBs. Figure 1B-2 is an example of these flash drives. The size and complexity of such systems ranges from portable USB jump drives to enterprise-class

---

reflective
/rɪˈflektɪv/
adj. 反射的
variation
/ˌveəriˈeɪʃn/
n. 变化
retrieve
/rɪˈtriːv/
vt. 检索
irregularity
/ɪˌregjʊˈlærətɪ/
n. 不规则
versatile
/ˈvɜːsətaɪl/
adj. 多用途的
Blu-ray
蓝光
violet
/ˈvaɪələt/
n. 紫色
spectrum
/ˈspektrəm/
n. 光谱;频谱
BD
蓝光盘
flash drives
闪存驱动器

Figure 1B-1　Blu-ray Disks

array-based memory systems. The high capacity of these portable units are easily connected to and disconnected from a computer. However, the **vulnerability** of their tiny storage **chambers** dictates that they are not as reliable as optical disks for truly long term applications.

Another application of flash technology is found in **SD**(**Secure Digital**) **memory cards**(or just SD Card), see Figure 1B-3. These provide up to four GBs of storage and are packaged in a plastic **rigged wafer** about the size a postage stamp, SDHC(High Capacity) memory cards can provide up to 32 GBs and the next generation SDXC(Extended Capacity) memory cards may exceed a TB. Given their compact physical size, these cards conveniently **slip into slots** of small electronic devices. Thus, they are ideal for digital cameras, smartphones, music players, car navigation systems, and a host of other electronic appliances.

Figure 1B-2　Flash Drive

Figure 1B-3　SD Card

- **Solid-state memory**

Unlike the disk and CD-ROM, a solid-state memory (see Figure 1B-4) doesn't need a read/write head, and the storage medium doesn't need to rotate to read and write data.

A solid-state memory store data by the switch state of transistors in the memory chip, because of the solid-state memory doesn't need a read/write head, and doesn't need to rotate, so it has the advantages of less **consumption** and strong **vibration resistance**. Due to the high cost, mechanical hard disks are still used in mass storage. But in small capacity, **ultra**-high speed and small electronic devices, solid-state memory has a huge advantage.

consumption
/kənˈsʌmpʃn/
n. 消耗；消费
vibration
/vaɪˈbreɪʃn/
n. 震动
resistance
/rɪˈzɪstəns/
n. 抵抗；反对
ultra
/ˈʌltrə/
adj. 超级；极端

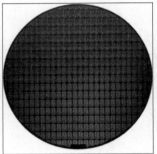

Figure 1B-4   Solid-state Memory

Exercises

Ⅰ. **Fill in the blanks with the information given in the text.**

1. Computers information is encoded as patterns of 0s and 1s which are called _____.

2. A string of _____ bits is called a byte.

3. To identify individual cells in a computer's main memory, each cell is assigned a unique "name", called its _____.

4. A computer's main memory is often called _____.

5. Most computers have additional memory devices called mass storage systems, including magnetic disks, CDs, DVDs, magnetic tapes, and _____.

6. _____ provides over five times the capacity of a DVD.

7. Another application of flash technology is found in _____ memory cards, which are ideal for digital cameras, _____, music players, car navigation systems, and a host of other electronic appliances.

Ⅱ. **Translate the following terms or phrases from English into Chinese.**

digit   binary
main memory   byte

address	RAM
mass storage	flash drive
compact disk	DVD
Blu-ray Disk	SD Card
solid-state memory	vibration resistance
ultra-high speed	

Section C 如何学习计算机专业英语

1. 为什么要学计算机专业英语

随着时代的发展,计算机已经渗透到人们工作与生活的方方面面,特别是当计算机技术与网络技术相结合,人类进入信息时代,计算机技术也在以前所未有的速度发展,人们接触到的计算机专业英语词汇不断增加。作为一个高技术行业,IT行业吸引着大量的从业人员。然而,计算机技术特别是软件技术更新的速度越来越快,而这些技术大多来源于英语国家,语言障碍将会严重影响到人们对新技术的理解与掌握。

编程本身就依赖于英语。目前,几乎所有的编程语句都使用英文,虽然某些开发工具的变量名和字段名可以支持中文,但不是主流。操作系统中的一些设置,如BIOS、注册表等也是英文,如果不能理解这些专业英语,就不能更好地发挥计算机的性能,使用好计算机。

软件开发中的技术文档和资料大都是英文,较先进的硬件设备的说明文档与手册也大多是英文,需要查阅的学科前沿资料也是英文居多。即便有翻译成中文的资料,但其中有不少内容晦涩难懂,译法混乱,脱离了原文的意思,使读者对原文产生误解。如果能直接阅读这些第一手的英文资料,不仅能及时掌握最新信息,也能透彻明白作者的意思。

计算机专业英语把当今最热门的两种学科计算机和英语相结合,如果能好好地掌握学习这两门学科,并加以结合,就等于拥有两样法宝在身上了。这也迫切要求我们不仅要学好计算机,更要学好这门与计算机沟通的语言工具——计算机专业英语。

2. 计算机专业英语与普通英语的区别

普通英语是广泛的,涵盖所有行业,主要使用通用、普遍的基础词汇。计算机专业英语主要指有关计算机方面的英语,广义上来说,其覆盖范围相当于IT业英语,涉及计算机软硬件、网络通信、人工智能、计算机安全等多个方面,计算机专业英语的专业词汇更多。计算机专业英语与普通英语之间有相同的地方,也有区别。

2.1 计算机专业英语与普通英语的共同点

在语音上,计算机专业英语使用的仍是普通英语的语音系统,并无自己独立的发音。

在词汇上,计算机专业英语中虽然有大量的专业技术词汇和术语,但其基础词汇都是普通英语中固有的。

在语法上,计算机专业英语虽然有明显的特点,但仍遵循普通英语的语法规则,并无独立的词法和句法系统。

2.2 计算机专业英语的特点

(1) 计算机专业英语的词汇特点

计算机专业英语的词汇是学习计算机专业英语的重要组成部分,但只靠死记硬背不仅效率不高,而且效果有限。要迅速扩大词汇量,加深对词汇的理解,需要了解计算机专业英语的词汇特点。

① 纯科技词汇。

这类词汇的特点是词义窄,拼写长,专业性强,出现频率不高。往往拼写越长的,词义越狭窄。

  例如：bandwidth——带宽
  hexadecimal——十六进制
  multitasking——多任务处理

② 通用科技词汇。

这类词汇的特点是词义窄,使用范围较广,出现频率较高。

  例如：frequency——频率
  density——密度

③ 半科技词汇。

这类词汇指的是在计算机专业英语中使用的普通词汇。它除了本身的基本词义外,在不同的专业中又有不同的含义。这类词汇的特点是词义多,使用范围广,出现频度高,较难掌握。

  例如：process——在普通英语中表示"处理、过程"等,而在计算机专业英语的操作系统领域则表示"进程"。
  bus——在普通英语中表示"公共汽车",而在计算机专业英语中表示"总线"。
  message——在普通英语中表示"消息、信息"等,而在通信中表示"报文"。

④ 抽象名词。

科技文章中偏爱使用概念准确的抽象名词,这类名词能够较好地表示过程、现象、特征和性质。它们大都由普通动词或形容词衍生而来。

  例如：available——availability(有效性)
  robust——robustness(健壮性)
  enhance——enhancement(增强)

⑤ 加前后缀构成的词。

这类词汇大量存在。在计算机专业英语中,词缀出现的频率远比其他文体的英语中要高。加前缀主要用来改变词义,但不改变词性;加后缀可能改变也可能不改变词义,但一定改变词性。

  例如：meter   *n*. 米——centi-——centimeter   *n*. 厘米
  compress   *vt*. 压缩——de-——decompress   *vt*. 解压缩
  simply   *adj*. 简单——-ify——simplify   *vt*. 简化
  use   *n*. 用途; *vt*. 使用——-less——useless   *adj*. 无用的

⑥ 缩略词。

在计算机专业英语中,缩略词的数量极为可观,并且新创造的缩略词仍在源源不断地补充进来,其数量之多,难以统计。很多缩略词使用得极其普遍,以至出现缩略词时不需要任何附加说明。

  例如：FTP 代表 file transfer protocol(文件传输协议)
  LCD 代表 liquid crystal display(液晶显示器)

GUI 代表 graphical user interface（图形用户界面）

（2）计算机专业英语的语法特点

计算机专业英语的语法特点主要表现为它具有很强的专业性。懂专业的人用起来得心应手，不懂专业的人用起来则困难重重。与普通英语相比，计算机专业英语更注重客观事实和真理，并且要求逻辑性强，用词规范，表达准确、精练，它的语法特点主要表现在以下几点。

① 客观性。

客观性是指表达的内容是客观的。为实现客观性，常用被动语态和一般现在时。即使用了主动语态，主语也常常是无生命的主语。就时态而言，因为专业资料内容涉及科学定义、定理或图表等，一般并没有特定的时间关系，所以在计算机专业英语中大部分使用一般现在时，而过去将来时、完成进行时很少出现。例如：

Multithreading is a program's capability to perform several tasks simultaneously.

译：多线程操作使程序能够同时执行多个任务。

② 准确性。

准确性是指意思表达要求准确。准确性主要表现在用词和造句上。为求精细地描述事物过程，所用句子都较长，有些甚至一段就是一个句子。长句可以反映客观事物中复杂的关系。例如：

A cloud must have a large and flexible resource pool to meet the consumer's needs, to provide the economies of scale, and to meet service-level requirements.

译：云必须具有大型的灵活资源池，以满足使用者的需求，提供规模化经济效益并满足服务级别要求。

Object-based storage is a way to store file data in the form of objects on flat address space based on its content and attributes rather than the name and location.

译：基于对象的存储是一种在单一地址空间上根据文件数据的内容和属性，而不是名称和位置，以对象的形式存储这些数据的方法。

③ 精练性。

精练性是指表达形式要求简洁、精练，希望用尽可能少的单词来清晰地表达原意，因而导致了非限定动词、名词化单词、词组及其简化形式的广泛使用。例如：

NAS(Network Attached Storage) is an IP-based, dedicated, high-performance file sharing and storage device.

译：NAS（网络连接存储）是一个基于 IP 网、专用、高性能的文件共享和存储设备。

3. 怎么学计算机专业英语

3.1 定目标

首先可以根据自己的实际英语水平和以后的工作需要进行分析，制定一个学习方向或目标。

一般而言，对于立志在泛 IT 业工作的学生来说，学习的目标应是掌握计算机专业英语中的常用术语和缩略语，了解计算机专业英语中常用的语法和习惯用法，能借助字典进行阅读，阅读速度在 70～90 词/min 等。

对于立志从事计算机研发的学生来说，学习的目标应定得高一些。需掌握大量计算机英语的专业术语，掌握计算机专业英语中常用的语法和习惯用法，阅读速度在 100 词/min 以上，能借助字典翻译专业技术图书，能使用英文编写简单的技术文档和程序注释等。

3.2 怎样学

学好计算机专业英语的关键在于短期的系统学习和长期的日常学习相结合。

短期的系统学习包括在校生参加计算机专业英语课程学习或是参加计算机专业英语培训班。通过使用计算机专业英语教材，将听老师讲解与自学配合，掌握计算机中的专业词汇和术语、翻译技巧等。

长期的日常学习更为重要。持之以恒，抓住各种机会学习，才能不断提高。可以选择下面的一些方式来学习：

- 自己计算机中的操作系统或是软件尽量安装英文版。
- 在硬件、软件的使用过程中如果遇到困难，尽量查阅英文帮助或技术资料。
- 上网时尽量多访问英文的技术论坛和网站。
- 坚持阅读英文计算机图书。对于不好读的地方，多作句法分析，或反复推敲，写出翻译的句子，体会西方人表达的思维方式。
- 坚持记录和复习遇到的计算机词汇、专业术语和缩略词。

其实，学习计算机专业英语并不是难事，了解了计算机专业英语的重要性，调动起学习的积极性和主动性，掌握学习方法，持之以恒，就一定能学好计算机专业英语。

Unit Two   Operating System and Computer Architecture

Section A   Operating System

An **OS** is a program that controls the execution of application programs and acts as an **interface** between applications and the computer hardware. It can be thought of as having three objectives:

- Convenience: An OS makes a computer more convenient to use.
- Efficiency: An OS allows the computer system resources to be used in an efficient manner.
- Ability to evolve: An OS should be constructed in such a way as to permit the effective development, testing, and introduction of new system functions without interfering with service.

There are four aspects of an OS.

OS
操作系统
interface
/ˈɪntəfeɪs/
n. 接口;界面

Ⅰ. OS as a User/Computer Interface

The hardware and software used in providing applications to a user can be viewed in a layered or **hierarchical** fashion, as depicted in Figure 2A-1.

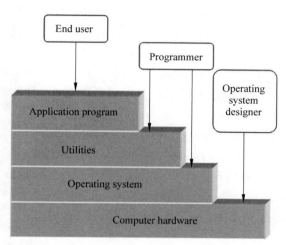

Figure 2A-1 Layers and Views of a Computer System

The user of those applications, the end user, generally is not concerned with the details of computer hardware. Thus, the end user views a computer system in terms of a set of applications. An application can be expressed in a programming language and is developed by an application programmer. If one were to develop an application program as a set of machine **instructions** that is completely responsible for controlling the computer hardware, one would be faced with an **overwhelmingly** complex **undertaking**.

To ease this **chore**, a set of system programs were provided. Some of these programs are referred to as **utilities**. These implement frequently used functions that assist in program creation, the management of files, and the control of **I/O** devices. A programmer will make use of these facilities in developing an application, and the application, while it is running, will **invoke** the utilities to perform certain functions.

The most important collection of system programs comprises the OS. The OS masks the details of the hardware from the programmer and provides the programmer with a convenient interface for using the system. It acts as **mediator**, making it easier for the programmer and for application programs to access and use those facilities and services.

II. OS as Resource Manager

A computer is a set of resources for the movement, storage, and processing of data and for the control of these functions. The OS is responsible for managing these resources. Like other computer programs, the OS provides instructions for the **processor**. The key difference is in the **intent** of the program. The OS directs the processor in the use of the other system resources and in the **timing** of its executing of other programs. But in order for the processor to do any of these things, it must cease executing the OS program and execute other programs. Thus, the OS **relinquishes** control for the processor to do some "useful" work and then resumes control long enough to prepare the processor to do the next piece of work.

Figure 2A-2 suggests the main resources that are managed by the OS. A portion of the OS is in main memory. This includes the **kernel**, which contains the most frequently used functions in the OS and, at a given time, other portions of the OS currently in use. The remainder of main memory contains user programs and data. The **allocation** of this resource (main memory) is controlled jointly by the OS and memory management hardware in the processor, as we shall see. The OS decides when an I/O device can be used by a program in **execution** and controls access to and use of files. The processor itself is a resource, and the OS must determine how much processor time is to be **devoted** to the execution of a particular user program. In the case of a multiple-processor system, this decision must **span** all of the processors.

III. Services Provided by OS

The OS provides a variety of facilities and services, such as **editors** and **debuggers**. Typically, these services are in the form of utility programs that, while not strictly part of the core of the OS.

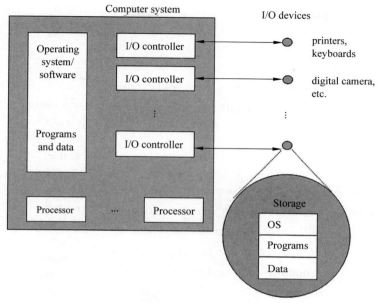

Figure 2A-2　OS as Resource Manager

- **Program execution**

A number of steps need to be performed to execute a program. Instructions and data must be loaded into main memory, I/O devices and files must be initialized, and other resources must be prepared. The OS handles these **scheduling** duties for the user.

- **Access to I/O devices**

Each I/O device requires its own peculiar set of instructions or control signals for operation. The OS provides a uniform interface that hides these details so that programmers can access such devices using simple reads and writes.

- **Controlled access to files**

For file access, the OS must reflect a detailed understanding of not only the nature of the I/O device but also the structure of the data contained in the files on the storage medium. In the case of a system with multiple users, the OS may provide protection mechanisms to control access to the files.

scheduling
/ˈʃedjuːlɪŋ/
n. 调度

- **System access**

For shared or public systems, the OS controls access to the system as a whole and to specific system resources. The access function must provide protection of resources and data from **unauthorized** users and must resolve **conflicts** for resource **contention**.

- **Error detection and response**

A variety of errors can occur while a computer system is running. These include internal and external hardware errors, such as a memory error, or a device failure or **malfunction**; and various software errors, such as division by zero, attempt to access forbidden memory location, and **inability** of the OS to grant the request of an application. In each case, the OS must provide a response that clears the error condition with the least impact on running applications.

- **Accounting**

A good OS will collect usage **statistics** for various resources and monitor performance parameters such as response time. On any system, this information is useful in anticipating the need for future enhancements and intoning the system to improve performance.

Ⅳ. Ease of Evolution of an Operating System

A major operating system will evolve over time for a number of reasons:

- **Hardware upgrades plus new types of hardware**

For example, early versions of UNIX[①] and the Macintosh[②]

unauthorized
/ˌʌnˈɔːθəraɪzd/
adj. 未经授权的

conflict
/ˈkɒnflɪkt/
n. 冲突

contention
/kənˈtenʃn/
n. 竞争

malfunction
/ˌmælˈfʌŋkʃn/
n. 故障

inability
/ˌɪnəˈbɪlətɪ/
n. 无能为力

statistics
/stəˈtɪstɪks/
n. 统计

---

① UNIX：操作系统，是一个强大的多用户、多任务操作系统，支持多种处理器架构，按照操作系统的分类，属于分时操作系统，最早由 Ken Thompson、Dennis Ritchie 和 Douglas McIlroy 于 1969 年在 AT&T 的贝尔实验室开发。

② Macintosh：麦金塔计算机，俗称 Mac 机，也称作苹果机或麦金托什机，是对苹果 PC 中一系列产品的称谓。首款 Mac 于 1984 年 1 月 24 日发布，是苹果公司继 Lisa 后第二款具备图形界面的 PC 产品。

operating system did not employ a **paging mechanism** because they were run on processors without paging hardware. Subsequent versions of these operating systems were modified to exploit paging capabilities. Also, the use of graphics terminals and page-mode terminals instead of **line-at-a-time** scroll mode terminals affects OS design. A graphics terminal typically allows the user to view several applications at the same time through "windows" on the screen. This requires more sophisticated support in the OS.

- **New services**

In response to user demand or in response to the needs of system managers, the OS expands to offer new services. For example, if it is found to be difficult to maintain good performance for users with existing tools, new measurement and control tools may be added to the OS.

- **Fixes**

Any OS has faults. These are discovered over the course of time and fixed are made. Of course, the fix may introduce new faults.

The need to change an OS regularly places certain requirement on its design. An obvious statement is that the system should be **modular** in construction, with clearly defined interfaces between the modules, and that it should be well documented. For large programs, such as the typical **contemporary** OS, what might be referred to as straightforward modularization is **inadequate**. That is, much more must be done than simply **partitioning** a program into modules.

Ⅴ. The Development of OS

Perhaps the best known example of an operating system is Windows[①], which is provided in numerous versions by Microsoft and widely used in the PC arena. Another well-established example

---

① Windows 是微软公司制作和研发的一套桌面操作系统,问世于 1985 年。Windows 不断升级,从架构的 16 位、32 位再到 64 位,系统版本从最初的 Windows 1.0 发展到 Windows 11 和 Server 服务器企业级操作系统。

is UNIX, which is a popular choice for larger computer systems as well as PCs. In fact, UNIX is the core of two other popular operating systems: Mac OS, which is the operating system provided by Apple for its range of Mac machines, and **Solaris**①, which was developed by Sun Microsystems② (now owned by ORACLE③). Still another example of an operating system found on both large and small machines is Linux④, which was originally developed noncommercially by computer **enthusiasts** and is now available through many commercial sources, including IBM.

In short, operating systems have grown from simple programs that retrieved and executed programs one at a time into complex systems that coordinate **timesharing**, maintain programs and data files in the machine's mass storage devices, and respond directly to requests from the computer's users. But the evolution of operating systems continues. The development of multiprocessor machines has led to operating systems that provide time-sharing/**multitasking** capabilities by assigning different tasks to different processors as well as by sharing the time of each single processor. These operating systems must **wrestle** with such problems as **load balancing** (dynamically allocating tasks to the various processors so that all processors are used efficiently) as well as scaling (breaking tasks into a number of subtasks **compatible** with the number of processors available).

Moreover, the advent of computer networks has led to the creation of software systems to coordinate the network's activities. Thus the field of networking in many ways is an extension of the subject of operating systems—the goal being to manage resources across many users on many machines rather than a single, **isolated** computer.

Still another direction of research in operating systems focuses on devices that are dedicated to specific tasks such as medical

---

① Solaris: Sun Microsystems 研发的计算机操作系统,被认为是 UNIX 操作系统的衍生版本之一。
② Sum Microsystems: 创建于 1982 年。主要产品是工作站及服务器。1992 年,Sun Microsystems 推出了市场上第一台多处理器台式机 SPARC station 10 system。
③ ORACLE: 甲骨文公司,全球最大的企业级软件公司,1977 年在 IBM 发表"关系数据库"的论文时,埃利森以此造出新数据库,名为甲骨文。
④ Linux: 一套免费使用和自由传播的类 UNIX 操作系统。

Solaris
/'səʊlərais:/
操作系统名

enthusiast
/ɪnˈθjuːziæst/
n. 热心者;热情者

timesharing
/ˈtaɪmʃeərɪŋ/
n. 分时(操作)

multitasking
/ˈmʌltɪɪtɑːskɪŋ/
n. 多任务处理

wrestle
/ˈresl/
v. 斗争

load balancing
负载平衡

compatible
/kəmˈpætəbl/
adj. 兼容的

isolated
/ˈaɪsəleɪtɪd/
adj. 孤立的

devices, **vehicle** electronics, home **appliances**, cell phones, or other hand-held computers. The computer systems found in these devices are known as **embedded systems**. Embedded operating systems are often expected to conserve battery power, meet demanding **real-time** deadlines, or operate continuously with little or no human **oversight**. Successes in this **endeavor** are marked by systems such as VxWorks①, developed by Wind River② and used in the Mars Exploration Rovers③ named Spirit and Opportunity; Windows CE④ also known as Pocket PC developed by Microsoft; and Palm OS developed by PalmSource, Inc., especially for use in hand-held devices.

Traditional time-sharing/multitasking systems give the **illusion** of executing many processes at once by switching rapidly between time slices faster than a human can perceive. Modern systems continue to multitask in this way, but in addition, the latest **multi-core** CPUs are **genuinely** capable of running two, four, or many more processes **simultaneously**. Unlike a group of single-core computers working together, a multi-core machine contains multiple independent processors (in this case called cores) that share the computer's **peripherals**, memory, and other resources. For a multi-core operating system, this means that the **dispatcher** and scheduler must consider which processes to execute on each core.

Ⅵ. Huawei HarmonyOS

Huawei⑤ HarmonyOS⑥ is an operating system officially

---

① VxWorks：1983年设计开发的一种嵌入式实时操作系统，它以良好的可靠性和卓越的实时性被广泛地应用在通信、军事、航空等高精尖技术及对实时性要求极高的领域中。

② Wind River：风河系统公司，是全球领先的嵌入式软件与服务商。

③ Mars Exploration Rovers：火星探测漫游者，美国国家航空航天局2003年的火星探测计划。这项计划的主要目的是将勇气号（Spirit）和机遇号（Opportunity）两辆火星车送往火星，对火星进行实地考察。

④ Windows CE：微软公司嵌入式、移动计算平台的基础，它是一个开放的、可升级的32位嵌入式操作系统。

⑤ Huawei：华为技术有限公司，成立于1987年，总部位于中国深圳，创始人为任正非。2021年《财富》公布世界500强榜，华为位于第44位，位列中国民营企业第一。

⑥ HarmonyOS：2019年华为发布的鸿蒙操作系统。"鸿蒙"的中文有开天辟地之意。

launched by Huawei at the Huawei Developer Conference (HDC①) held in Dongguan on August 9, 2019, as part of its broader push to solve China's problem of lacking homegrown operating systems for fundamental digital technologies.

HarmonyOS is a new distributed operating system oriented to the whole scene, creating a super virtual terminal interconnected world, connecting people, equipment and scene **organically**, realizing rapid discovery, rapid connection, hardware mutual assistance and resource sharing of various **intelligent terminals** that consumers come into contact with in the whole scene of life.

On September 10, 2020, Huawei said that its **in-house operating system** HarmonyOS would be used in smartphones next year, marking a breakthrough in Chinese companies' efforts to commercialize self-developed operating systems and to build their own globally competitive software **ecosystems**. The number of developers of HMS② stood at over 1.8 million and the mobile applications integrated with HMS so far have exceeded 96,000.

HarmonyOS is already used in Huawei's smart TV products. With the upgrade of the system, and will be used in smartwatches, personal computers and other **Internet of Things** (IoT) devices later.

Huawei has also promised to make HarmonyOS **open source**, which means anyone can freely examine the system specification to make sure there's no problem. The code for small Internet of Things devices with 128 megabytes or less storage is available now. The code for larger devices will be freely published in April 2021, and the remaining code will be available for download by October 2021.

organically
/ɔːˈɡænɪkli/
adv. 有机地
intelligent terminal
智能终端
in-house operating system
自主研发的操作系统
ecosystem
/ˈiːkəʊsɪstəm/
n. 生态系统

Internet of Things
物联网
open source
开源

Exercises

Ⅰ. Fill in the blanks with the information given in the text.

1. An OS is a program that controls the execution of application programs and acts as an _____ between applications and the computer hardware.

---

① Huawei Developer Conference (HDC):华为开发者大会。
② HMS:Huawei Mobile Services,华为移动服务,是华为为其设备生态系统提供的一套应用程序和服务,包含华为账号、应用内支付、华为推送服务、华为云盘服务、华为广告服务、消息服务、付费下载服务、快应用等服务。

2. The OS provides a variety of facilities and services, such as _____ and _____.

3. Instructions and data must be loaded into _____ before being executed.

4. The OS provides a _____ interface that hides these details so that programmers can access such devices using simple reads and writes.

5. In the case of a system with _____ users, the OS may provide protection mechanisms to control access to the files.

6. A variety of errors can occur while a computer system is running. These include internal and external _____ errors, and various software errors.

7. A good OS will collect usage statistics for various resources and performance parameters such as response time.

8. The use of _____ terminals and _____ terminals instead of line-at-a-time scroll mode terminals affects OS design.

9. Any OS has _____. These are discovered over the course of time and fixed are made.

10. For large programs, such as the typical contemporary OS, what might be referred to as straightforward _____ is inadequate.

Ⅱ. **Translate the following terms or phrases from English into Chinese.**

Operating System, OS	execution
application program	interface
hierarchical	programmer
instruction	utility
editor	debugger
I/O	scheduling
disk drive	internal
time-sharing	multitasking
load balancing	embedded system
real-time	multi-core
dispatcher	scheduler
parameter	multi-user
storage	allocation
processor	kernel
version	page
modular	HarmonyOS
intelligent terminal	in-house operating system
self-developed operating system	software ecosystem
open source	

Ⅲ. **Translate the following passage from English into Chinese.**

Linux

For the computer enthusiast who wants to experiment with the internal components of an operating system, there is Linux. Linux is an operating system originally designed by Linus Torvalds while a student at the University of Helsinki. It is a nonproprietary product and available, along with its source code and documentation, without charge. Because it is freely available in source code form, it has become popular among computer hobbyists, students of operating systems, and programmers in general. Moreover, Linux is recognized as one of the more reliable operating systems available today. For this reason, several companies now package and market versions of Linux in an easily useable form and these products are now challenging the long-established commercial operating systems on the market. You can learn more about Linux from the Web site at http://www.linux.org.

Section B Computer Architecture

Ⅰ. CPU

The circuitry in a computer that controls the manipulation of data is called the central processing unit, or **CPU**. In the machines of the mid-twentieth century, CPUs were large units comprised of perhaps several **racks** of electronic circuitry that reflected the significance of the unit. However, technology has **shrunk** these devices drastically. The CPUs found in today's desktop computers and notebooks are packaged as small flat **squares** (approximately two inches by two inches) whose connecting pins plug into a socket mounted on the machine's main circuit board(called the **motherboard**). In smartphones, mini-notebooks, and other Mobile Internet Devices(MID), CPU's are around half the size of a postage stamp. Due to their small size, these processors are called microprocessors.

A CPU consists of three parts: the **arithmetic**/logic unit, which contains the circuitry that performs operations on data; the control unit, which contains the circuitry for coordinating the machine's activities; and the **register** unit, which contains data storage cells, called registers, that are used for temporary storage of information within the CPU.

Ⅱ. The Stored-Program Concept

A breakthrough came with the realization that a program, just like data, can be encoded and stored in main memory presented by

CPU
中央处理器
rack
/ræk/
n. 机架;货架
shrink
/ʃrɪŋk/
v. 收缩;缩水
square
/skweə/
n. 正方形;平方
motherboard
/ˈmʌðəbɔːd/
n. 主板
MID
移动互联设备
arithmetic
/əˈrɪθmətɪk/
n. 算术;计算
register
/ˈredʒɪstə/
n. 寄存器

John von Neumann①. If the control unit is designed to **extract** the program from memory, decode the instructions, and execute them, the program that the machine follows can be changed merely by changing the contents of the computer's memory instead of rewiring the CPU.

The idea of storing a computer's program in its main memory is called the **stored-program** concept and has become the standard approach used today. What made it difficult originally was that everyone thought of programs and data as different **entities**: Data were stored in memory; programs were part of the CPU. The result was a prime example of not seeing the forest for the trees. Indeed, part of the excitement of the science is that new insights are constantly opening doors to new theories and applications.

III. Other Architectures

To broaden our perspective, let us consider some alternatives to the traditional machine architecture we have discussed so far.

- **Pipelining**

Electric pulses travel through a wire no faster than the speed of light. Since light travels approximately 1 foot in a **nanosecond**, it requires at least 2 nanoseconds for the CPU to fetch an instruction from a memory cell that is 1 foot away. Consequently, to fetch and execute an instruction in such a machine requires several nanoseconds—which means that increasing the execution speed of a machine ultimately becomes a miniaturization problem.

However, increasing execution speed is not the only way to improve a computer's performance. The real goal is to improve the machine's **throughput**, which refers to the total amount of work the machine can accomplish in a given amount of time.

An example of how a computer's throughput can be increased without requiring an increase in execution speed involves pipelining, which is the technique of allowing the steps in the machine cycle to **overlap**. In particular, while one instruction is being executed, the next instruction can be fetched, which means

---

① 约翰·冯·诺依曼(1903—1957),美籍匈牙利数学家,提出了"存储程序"的思想。

extract
/ˈekstrækt/
vt. 提取

stored-program
存储程序

entity
/ˈentɪti/
n. 实体

pipelining
流水线
nanosecond
/ˈnænəʊˌsekənd/
n. 纳秒;
十亿分之一秒

throughput
/ˈθruːpʊt/
n. 生产量;吞吐量

overlap
/ˌəʊvəˈlæp/
v. 重叠;相交

that more than one instruction can be in "the **pipe**" at any one time, each at a different stage of being processed. In turn, the total throughput of the machine is increased even though the time required to fetch and execute each individual instruction remains the same.

Modern machine designs push the pipelining concept beyond our simple example. They are often capable of fetching several instructions at the same time and actually executing more than one instruction at a time when those instructions do not rely on each other.

- **Multiprocessor Machines**

Pipelining can be viewed as a first step toward **parallel** processing, which is the performance of several activities at the same time. However, true parallel processing requires more than one processing unit, resulting in computers known as multiprocessor machines.

A variety of computers today are designed with this idea in mind. One strategy is to attach several processing units, each **resembling** the CPU in a single processor machine, to the same main memory. In this configuration, the processors can proceed independently yet coordinate their efforts by leaving messages to one another in the common memory cells. For instance, when one processor is faced with a large task, it can store a program for part of that task in the common memory and then request another processor to execute it. The result is a machine in which different instruction sequences are performed on different sets of data, which is called a **MIMD**(multiple-instruction stream, multiple-data stream) architecture, as opposed to the more traditional **SISD** (single-instruction stream, single-data stream) architecture.

A variation of multiple-processor architecture is to link the processors together so that they execute the same sequence of instructions in **unison**, each with its own set of data. This leads to a **SIMD**(single-instruction stream, multiple-data stream) architecture.

Another approach to parallel processing is to construct large computers as **conglomerates** of smaller machines, each with its own memory and CPU. Within such an architecture, each of the small

pipe
/paɪp/
n. 管子；管道

parallel
/ˈpærəlel/
adj. 并行的

resemble
/rɪˈzembl/
v. 类似；像

MIMD
多指令多数据流
SISD
单指令单数据流

unison
/ˈjuːnɪzn/
n. 一致；协调
SIMD
单指令多数据流
conglomerate
/kənˈglɒmərɪt/
n. 集团；聚块

machines is **coupled** to its neighbors so that tasks assigned to the whole system can be divided among the individual machines. Thus if a task assigned to one of the internal machines can be broken into independent subtasks, that machine can ask its neighbors to perform these subtasks concurrently.

couple
/ˈkʌpl/
v. 连接

Exercises

Ⅰ. Fill in the blanks with the information given in the text.

1. The circuitry in a computer that controls the manipulation of data is called the _____.

2. A CPU consists of three parts: the arithmetic/logic unit, _____, and the register unit.

3. The idea of storing a computer's program in its main memory is called the _____.

4. The real goal is to improve the machine's _____, which refers to the total amount of work the machine can accomplish in a given amount of time.

5. _____ can be viewed as a first step toward parallel processing, which is the performance of several activities at the same time.

Ⅱ. Translate the following terms or phrases from English into Chinese.

central processing unit, CPU   microprocessor
motherboard   stored-program concept
register   pipelining
throughput   MIMD
parallel processing   SIMD

Section C 计算机专业英语词汇

学习计算机专业英语的关键点与难点之一就是记忆计算机词汇、专业术语。具备一定的词汇量,对于提高阅读能力和水平都有一定的帮助。但是,机械地逐一记忆会花费大量的时间。而且,计算机行业不断涌现一些新构造的单词,在字典中也查不到。因此,必须了解计算机专业英语中词汇的特色,学习计算机专业英语的构词法。

计算机专业英语中的词汇构成主要是通过词义转移、引申、词类转换、词语的复合、缩略、词语的裁剪、拼合和词缀构词法等方式来形成计算机专业英语特殊的词汇体系。

1. 词义转移、引申

许多计算机专业英语词汇是从普通英语中转移过来的,词义转移或引申是构建计算机英语词汇的一种常见方法。这一类旧词有了新的释义,被赋予新义的词汇拼写不变。在学习中特别要注意,如果按照普通英语的词义去理解,句子就错了,因此需要特别记住这一类词。例如:

mouse 老鼠→鼠标
driver 司机→驱动程序
call 电话→调用
explorer 探索者→浏览器
bridge 桥→网桥
page 页→网页

2. 词语的复合

复合法是指把两个或两个以上独立的词结合在一起构成一个新词的方法。用复合法构成的词称为复合词。复合词的词义通常可以根据其组合成分推知,但要注意其词义往往并不是其组合成分各自意义的简单相加。从书写特征看,复合词有连写、用连字符、分写三种形式。例如:

back＋up→backup 备份
byte＋code→bytecode 字节码
back＋space→backspace 退格
gate＋way→gateway 网关
object＋oriented→object-oriented 面向对象的
sign＋on→sign-on 登录
hard＋disk→hard disk 硬盘
read＋me→read me 自述

3. 缩略法

缩略法指通过省略或简化词的音节来构成一个新词。用缩略法构成的词称为缩略词。缩略词主要有以下4种类型。

（1）首字母缩略词

用一个词组中的各个词语或各主要词的第一个字母或前几个字母组成。例如：

CPU 代表 central processing unit（中央处理器），取三个词的第一个字母组成。

ASCII 代表 American Standard Code for Information Interchange（美国信息交换标准码），取 for 以外的各词的首字母组成（在首字母缩写词中，of、for、and 等结构词或虚词一般都不表现，实词或实义词一般都有体现）。

DBMS 代表 database management system（数据库管理系统），取第一个词的两个字母和后两个词的首字母组成。

B2B 代表 Business to Business（企业对企业），用发音相同的数字代替单词。

（2）首字母拼音词

首字母拼音词与首字母缩略词的构词方式相同，区别在于，首字母拼音词成词后能拼读成一个词。例如：

ROM 代表 read-only memory（只读存储器），读作/rɔm/

RAM 代表 random-access memory（随机存储器），读作/ræm/

（3）截短词

截短词是通过截除原词的某些音节而构成的词，多用于口语或非正式场合。例如：

information→info（信息，读作/ˈɪnfəʊ/）

application→app（应用程序，读作/æp/）

commercial→com（作为网址的组成部分，表示商业机构）

（4）拼缀词

拼缀词是将原有的两个词中的其中一个或两个加以裁剪，取其中的首部或尾部，然后连接成一个新词。例如：

bit（二进制位，位，比特）＝binary（二进制的）＋digit（数字）

modem（调制解调器）＝modulator（调制器）＋demodulator（解调器）

malware（恶意软件）＝malicious（恶意的）＋software（软件）

4. 前缀、词根及后缀

计算机专业英语中有一大部分的单词是一个词根，前面加上前缀，或是后面加上后缀，而形成新的单词，新形成的单词或多或少都与共同的词根有相关联的意思。只要我们认识基本的词根，知道前缀、后缀表示的意思，通常就能理解新词的意思了。例如：

contract＝con＋tract：con 表示"共同，一起"，tract 表示"拉"，本词意思为"缩约词"。

distract＝dis＋tract：dis 表示"分开"，tract 表示"拉"，本词意思为"分散"。

下面列举一些常用的词缀。

4.1 前缀

前缀由一个或多个字母组成，放在词根或单词之前，以组成一个新词。每一个前缀都具有一定的含义，前缀一般不会导致词性的改变，而只改变原来单词的含义。

a-无，不：　　　　　　　　　aperiodic 非周期的；acentric 无中心的

ab-离开,脱离:	abnormal 不正常的
anti-反对:	antivirus 防病毒;antimagnetic 防磁的
auto-自动:	autosave 自动保存;autorotation 自动旋转
bi-双,两,二:	bidirectional 双向的;bicoloured 双色的
bio-生物,生命:	biochip 生物芯片;biology 生物学
by-边,副:	by-effect 副作用
centi-百分之一:	centimeter 厘米
co-共同,一起:	cooperation 合作;coaxial 同轴的
con-与,共,合:	connectivity 连通性;concentrated 集中的
com-与,共,合:	compress 压缩
counter-反,对应:	counteraction 反作用;counterclockwise 逆时针方向
de-相反,取消:	decompress 解压缩
dis-否定,相反:	disassemble 反汇编;discharge 放电
en-使:	enable 使能够;enlarge 扩大
ex-向外,超出:	export 导出;exchange 交换
extra-超出:	extraordinary 异常的
in-内,向内:	input 输入;inlet 入口
in-不,非,无:	incorrect 不正确的;incompatible 不可兼容的
注意:在 b,m,p 前加 im,在 l 前加 il。impossible 不可能的;illogical 不合逻辑的。	
inter-互相,在……之间:	internet 互联网;interact 互相作用
intra-在内,内部:	intranet 内联网
kilo-千:	kilobyte 千字节;kilogram 千克
macro-大的,宏观的:	macrostructure 宏观结构
mal-不良,不当:	malfunction 发生故障
micro-微小的:	microcomputer 微型计算机;microswitch 微型开关
mid-中间:	midterm 期中
mini-极小的:	minicomputer 小型计算机
mis-错,误:	misoperation 误操作
mono-单,一:	monobus 单总线;monotone 单音
multi-多:	multimedia 多媒体;multithread 多线程
non-非,无,不:	nonvolatile 非易失性的;nonconductor 绝绝缘体
out-在外,超过:	output 输出
over-过分,额外:	overload 过载;overrun 溢出
poly-多:	polymorphism 多态性;polycrystal 多晶体
post-后:	postwar 战后
pre-前,预先:	precompiler 预编译程序
pseudo-伪,假:	pseudocode 伪代码
re-再,重新:	reformat 重新格式化;rewrite 改写

semi-半：　　　　　　　　　　semiconductor 半导体；semidiameter 半径
sub-下面，次于：　　　　　　subroutine 子例程；subdirectory 子目录
super-超：　　　　　　　　　supercomputer 超级计算机；superpower 超功率
tele-远，电：　　　　　　　　telecommunication 电信；telephone 电话
ultra-超过：　　　　　　　　ultrared 红外线的；ultrastability 超稳定性
un-不，非：　　　　　　　　unauthorized 未经授权的；uninstall 卸载
under-在……下，不足：　　　underground 地下的；undergo 经历

4.2 后缀

后缀由一个或多个字母组成，位于词根或单词之后，组成一个新词。后缀不仅改变原单词的含义，还可以改变单词的词性，实现词类转变。因此，给单词添加后缀是构成新词，尤其是不同类型词类的常用方法。根据加后缀构成单词的词性，可以把后缀分为名词后缀、动词后缀、形容词后缀及副词后缀等。

4.2.1 常见的名词后缀

（1）表示人或物

-an/-ian/-arian：　　　　　technician 技师
-ant/-ent：　　　　　　　　assistant 助手；agent 代理人；coolant 冷却剂
-ee：　　　　　　　　　　　trainee 受培训者；employee 雇员，职工
-er/-or/-ar：　　　　　　　programmer 程序员；processor 处理器；hacker 黑客
-ese：　　　　　　　　　　computerese 计算机行话；Chinese 中国人
-ist：　　　　　　　　　　　specialist 专家；typist 打字员

（2）表示性质、状态、动作、过程等，构成抽象名词

-ability/-ibility：　　　　　scalability 可扩展性；interoperability 互操作性
-age：　　　　　　　　　　storage 存储；message 信息，报文
-ance/-ence：　　　　　　　inheritance 继承；conference 会议
-ancy/-ency：　　　　　　　inconsistency 不一致性；dependency 依赖性
-cy：　　　　　　　　　　　accuracy 准确性；captaincy 船长职位
-hood：　　　　　　　　　　likelihood 可能性；falsehood 虚假，谬误
-ion/-ation/-ition：　　　　computation 计算；specification 说明
-ity/-ty：　　　　　　　　　functionality 功能性；security 安全
-ment：　　　　　　　　　　enhancement 增强性；equipment 设备
-ness：　　　　　　　　　　robustness 健壮性；firmness 结实
-ship：　　　　　　　　　　partnership 伙伴关系；ownership 所有权
-sion/-ssion：　　　　　　　transmission 传输；decision 决定
-ure：　　　　　　　　　　　erasure 擦除；failure 失败

4.2.2 常见的形容词后缀

（1）表示具有……性质的

-al：　　　　　　　　　　　computational 计算的；digital 数字的；decimal 小数的

-ant/-ent： abundant 丰富的；vacant 空的；patent 专利的
-ary： customary 通常的；elementary 初级的
-ate： accurate 精确的
-ic/-ical： graphical 图形的；elastic 弹性的；electronic 电子的
-ed： skilled 熟练的；complicated 复杂的

（2）表示充满或缺乏
-free： jumper-free 无跳线的；salt-free 无盐的
-ful： powerful 有力的；useful 有用的
-less： wireless 无线的；useless 无用的
-ous/-ious： dangerous 危险的；enormous 巨大的
-y： noisy 有噪声的；healthy 健康的

（3）表示类似
-ish： tallish 有点高；youngish 有点年轻
-like： starlike 星形的；worm-like 像蠕虫

（4）表示能……的，易于……的
-able： executable 可执行的；adjustable 可调整的
-ible： compatible 兼容的；sensible 可感知的

（5）表示有……性质的，属于……的
-ive： active 积极的；productive 生产的

4.2.3 常见的动词后缀

（1）表示处理、使、成为……
-ate： eliminate 排除；simulate 模拟
-ize/-ise： computerize 计算机化
-en： soften 软化；lengthen 加长

（2）表示使……化
-ify： simplify 简化；magnify 放大

（3）表示方向、方式、状态
-wise： clockwise 顺时针方向；likewise 同样的

4.2.4 常见的副词后缀

-ly： programmatically 用编程方法
-ward/-wards： downwards 向下地；outwards 向外地

Unit Three Software Engineering

Section A Software Engineering Methodologies

Software development is an engineering process. The goal of researchers in software engineering is to find principles that guide the software development process and lead to efficient, reliable software products.

Ⅰ. Waterfall Model

The waterfall model (see Figure 3A-1) is a sequential design process, used in software development processes, in which progress is seen as flowing steadily downwards through the phases of requirements analysis, design, implementation, testing, and maintenance.

The waterfall development model originates in the manufacturing and construction industries; highly structured physical environments in which after-the-fact changes are **prohibitively** costly. The first known presentation describing use of similar phases in software engineering was held by Herbert D. Benington at **Symposium** on advanced programming methods for digital computers① on 29 June 1956.

---

① 赫伯特·D.贝宁顿在数字计算机高级编程方法研讨会上第一次在软件工程中提出了类似的阶段。

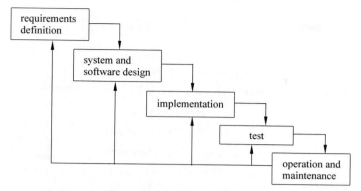

Figure 3A-1 Waterfall Model

II. Incremental Model

In recent years, software engineering techniques have changed to reflect the **contradiction** between the highly structured environment **dictated** by the waterfall model and the "**free-wheeling**", **trial-and-error** process that is often vital to creative problem solving. This is illustrated by the emergence of the **incremental model** (see Figure 3A-2) for software development. Following this model, the desired software system is constructed in increments—the first being a simplified version of the final product with limited functionality. Once this version has been tested and perhaps **evaluated** by the future user, more features are added and tested in an incremental manner until the system is complete.

contradiction
/ˌkɒntrəˈdɪkʃən/
n. 矛盾
dictate
/dɪkˈteɪt/
v. 控制；命令
free-wheeling
/ˌfriːˈwiːlɪŋ/
单向转动
trial-and-error
试错（法）；
不断摸索
incremental model
增量模型
evaluate
/ɪˈvæljueɪt/
v. 评估

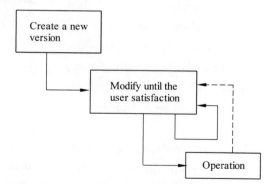

Figure 3A-2 Incremental Model

Ⅲ. Iterative Model

Another model that represents the shift away from strict adherence to the waterfall model is the **iterative model**(see Figure 3A-3), which is similar to, and in fact sometimes equated with, the incremental model, although the two are distinct. Whereas the incremental model carries the notion of extending each **preliminary** version of a product into a larger version, the iterative model **encompasses** the concept of **refining** each version. In reality, the incremental model involves an underlying iterative process, and the iterative model may incrementally add features.

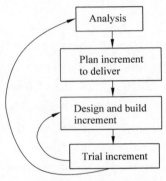

Figure 3A-3　Iterative Model

A significant example of iterative techniques is the **rational unified process**(**RUP**)(see Figure 3A-4) that was created by the Rational① Software Corporation, which is now a division of IBM. Generally, RUP is a development plan, which specifies the general process of developing a software product. Precisely, definition that states, which activities are to be performed, by which person, acting in which role; in which order the activities will be performed, and which products will be developed and how to evaluate them.

Incremental and iterative models sometimes make use of the trend in software development toward **prototyping** in which incomplete versions of the proposed system, called prototypes, are

---

① Rational：提供基于业界开放标准的工具、最佳方案和服务，用于开发商业应用和构建软件产品及系统，包括移动电话和医疗系统等设备使用的嵌入式软件。

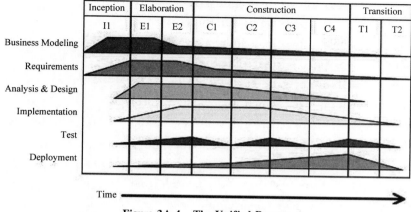

Figure 3A-4   The Unified Process

built and evaluated.

A less formal **incarnation** of incremental and iterative ideas that has been used for years by computer enthusiasts is known as **open-source development**. This is the means by which much of today's free software is produced. Perhaps the most prominent example is the Linux operating system whose open-source development was originally led by Linus Torvalds[①].

Ⅳ. Agile Methods

Perhaps the most **pronounced** shift from the waterfall model is represented by the agile methods, each of which proposes early and quick implementation on an incremental basis, **responsiveness** to changing requirements, and a reduced emphasis on **rigorous** requirements analysis and design. One example of an agile method is **extreme programming(XP)**.

XP was created by Kent Beck[②] during his work on the Chrysler Comprehensive Compensation System payroll project in the 1990's. In 1999, Beck defined XP as a lightweight methodology for the small-to-medium sized teams developing software in the face of vague or rapidly changing requirements. In terms of this definition,

---

① Linus Torvalds：林纳斯·托瓦尔兹(1969—)，芬兰人，著名的计算机程序员、黑客。Linux 内核的发明人及该计划的合作者。
② Kent Beck：肯特·贝克(1961—)，美国人，软件开发大师，是最早研究软件开发模式和重构的人之一，是敏捷方法的开创者之一，对当今世界的软件开发影响深远。

XP could be described in four ways. Firstly, it is lightweight. In XP you only do what you need to do to create value for the customer. Secondly, it is a methodology, which based on addressing obstacle in software development. It does not address some specific issues such as financial justification of projects or sales. Thirdly, it is implemented by a small or medium size, which makes the project team much more flexible. In this kind of team, everyone is elite and has his clear responsibility. Last but not least, XP adapts to rapidly changing requirements. In the modem business world requirements need to change to adapt to rapid shifts, so that the vital role that XP plays on the operating system cannot be over-emphasized.

Figure 3A-5 The Logo of XP

The contrasts depicted by comparing the waterfall model and XP reveal the **breadth** of methodologies that are being applied to the software development process in the hopes of finding better ways to construct reliable software in an efficient manner. Research in the field is an ongoing process. Progress is being made, but much work remains to be done.

breadth
/bredθ/
n. 宽度;广泛

Exercises

Ⅰ. **Fill in the blanks with the information given in the text.**

1. The goal of researchers in _____ is to find principles that guide the software development process and lead to efficient, reliable software products.

2. Early approaches to software engineering insisted on performing requirements analysis, _____, implementation, _____ and maintenance in a strictly sequential manner.

3. Following incremental model, the desired software system is constructed in _____ — the first being a simplified version of the final product with limited functionality.

4. A significant example of iterative techniques is the _____, which specifies the general process of developing a software product.

5. The most prominent example is the _____ operating system whose open-source development was originally led by Linus Torvalds.

6. Following the _____ model, software is developed by a team of less than a dozen individuals working in a communal work space where they freely share ideas and assist each other in the development project.

II. Translate the following terms or phrases from English into Chinese.

software engineering	requirements analysis
testing	maintenance
software engineer	requirements specification
waterfall model	incremental model
iterative model	rational unified process(RUP)
software life cycle	open-source
agile method	extreme programming(XP)

III. Translate the following passage from English into Chinese.

Software Ownership

Legal efforts to provide such ownership fall under the category of intellectual property law, much of which is based on the well-established principles of copyright and patent law. Indeed, the purpose of a copyright or patent is to allow the developer of a "product" to release that product to intended parties while protecting his or her ownership rights. As such, the developer of a product (whether an individual or a corporation) will assert his or her ownership by including a copyright statement in all produced works; including requirement specifications, design documents, source code, test plans, and in some visible place within the final product. A copyright notice clearly identifies ownership, the personnel authorized to use the work, and other restrictions. Furthermore, the rights of the developer are formally expressed in legal terms in a software license.

Section B Exploratory Testing

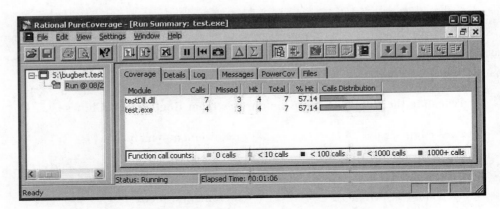

Software testing is a process where a software tester/team runs a program or a system to find **bugs** or defects, to maintain the correctness and reliability of a program. Software testing also validates and verifies the program to check if the business and technical requirements are met, and is working as expected.

Ⅰ. What is Exploratory Testing?

Exploratory testing—a term coined by Cem Kaner[①]—is a software testing method that implements learning, test design, and test execution at the same time simply because you explore while testing. It really doesn't follow a set of procedures, scripts or standards that is why it is mistakenly known as the "do it yourself" or careless activity, which is in fact, this approach describes an intellectual testing. During testing, the tester knows how the software really works, develop test design based on learning and experience, and the control to perform the entire testing stage. Because of the freedom that the exploratory testing offers, it solely depends on the tester's skill to find bugs, create **test cases**, and the responsibility to deliver the project optimizing the quality and value of the software.

bug
/bʌg/
n.(软件)错误；漏洞

exploratory testing
探索性测试

test case
测试用例

---

① Cem Kaner：卡尼尔，美国人，佛罗里达理工大学教授，在软件测试领域很出名。

II. Situational Behavior of Exploratory Testing

Unlike **scripted testing** that follow procedures, the exploratory testing is extensive in such a way that it is a **trial and error** situation. Performing this approach requires keen observation of the behavior, thorough investigation, critical thinking/**tactics** in finding bugs, analyze the possible issues, and evaluate the entire software.

A situational example for this is playing an online game. At first, the user has to know what it is all about, the objective, and how to play the game. While playing, the user now identifies the environment and functionalities of the game. The user discovers then the encountered difficulties and foresees potential problems on each level. At the same time, the user thinks strategies/solutions and also executes them as well and be able to complete and achieve the objective of the game.

Therefore, relating the situational example to exploratory testing, the following are the keys to remember:

(a) Identify the project's purpose.

(b) Determine the requirements and functionalities of the software.

(c) Identify the limitation of the software.

(d) Test the functionalities thoroughly based on the requirements.

(e) Create test cases that verify the consistency of the software.

scripted testing
脚本测试
trial and error
反复试验
tactic
/ˈtæktɪk/
n. 战术

III. Advantages

- Less preparation before the implementing the software testing because it doesn't require writing repeatable test procedures.
- Testing, designing and executing the software simultaneously saves time.
- Report many issues caused by incomplete or wrong documentation.

IV. Disadvantages

- Test designs created can't be reviewed prior to actual testing which might produce possible issues.
- Minor issues are poorly detected.
- Difficulty to perform the exact manner especially for new found bugs.

V. Conclusion

Exploratory Testing is necessary to practice because it enhances the tester's mind and strategic skills through exploration and experience. Therefore, the more the tester knows the keys in using this method, the better the testing, and the quality of software project will be.

Exercises

I. Fill in the blanks with the information given in the text.

1. Software testing is a process where a software tester/team runs a _____ or a system to find _____ or defects, to maintain the correctness and reliability of a program.

2. Exploratory testing is a software testing method that implements learning, test design, and test execution at the same time simply because you _____ while testing.

3. Exploratory testing solely depends on the tester's _____ to find bugs, create test cases.

4. Exploratory testing has less _____ before the implementing the software testing.

II. Translate the following terms or phrases from English into Chinese.

software testing   exploratory testing
scripted testing   trial and error
bug   test case

Section C 计算机专业英语翻译

1. 翻译的要点和标准

翻译就是把一种语言表达的信息用另一种语言表达出来。这里主要介绍把计算机英文资料用中文加以表达的方法和技巧。翻译不仅要求对英文原文有透彻、正确的理解,而且要求用符合中文语言习惯的适当表达方式进行再表述;不仅需要计算机专业知识,而且要求掌握必要的文化与背景知识。

翻译有翻译的标准。在我国,关于翻译标准问题影响最大的,当推严复的"信、达、雅"。"信"就是要忠实于原文,准确译出原意和原味。"达"就是要达意,语言要适当,表达要能让人懂。"雅"就是要文采,译文要给人以赏心悦目的美感。就计算机专业英文翻译而言,主要是达到前两条标准。

翻译的方法可分为"直译"和"意译"两大类。对于技术型较强的资料,译者不熟悉、新的、前沿性的内容,应"直译",力求准确,保证其与严谨性;对于广告性资料,译者十分了解、熟练掌握的内容,不妨"意译",可以使译文更易读。在具体的翻译实践中,应尽量兼顾这两方面。

2. 英语科技文章在句子结构上的特点

除了掌握计算机专业知识,熟记大量的计算机专业英语词汇,还需要了解计算机专业英语句子在结构上的特点。归纳为以下"八多":

- 陈述句多(主系表结构占较大比重)。
- 祈使句多(科技文章可操作性强,有时无须指明主语)。
- 被动句多(科技文章侧重叙事推理,强调客观准确)。
- 复合句多(并列关系、多种主从关系和非谓语动词构成的长句多)。
- 虚拟语气多(因科技文章常涉及各种条件)。
- 三种基本时态多(即一般现在时、一般过去时、现在完成时多)。
- as 引出的句式多(如 as shown in figure2,as stated above 等)。
- "It be+形容词或分词+that …"句型多。

3. 计算机专业英语的翻译技巧

翻译有一定的规律和方法可循。但是,翻译是一门需要大量实践的技能,只有通过自己的实践,并不断总结经验,才能提高翻译水平。下面讨论计算机专业英语翻译的一些常用方法,供大家在翻译实践中适当运用。

3.1 词义选择

每个英语单词都有一个或几个具体的含义,这就需要在翻译时选择一个恰当的词义。计算机英语是一种专业英语,其专业词汇大多是英语中已有的单词,只不过通过借用或引申原义等手段变成了计算机专业词汇。在具体的翻译中,可以根据上下文以及词的类别、词性、搭配等方面来判断和选择词义。

例句1：

The two main types of storage devices are disk drives and memory. 存储设备主要分两种类型：外存和内存。

在计算机专业英语中，storage 通常翻译成"存储"，memory 有"记忆、存储器、内存、回忆"等含义，但是根据上下文内容，此处介绍存储设备，那么 memory 应该翻译成"内存"，disk drives 本身有"磁盘驱动器"的意思，既然 memory 翻译成"内存"了，那么根据计算机知识翻译的习惯，此处的 disk drives 就翻译成"外存"。

例句2：

I have to install with each new PC：Office Suite, including E-mail client…我给每台新计算机必须安装的应用程序有：Office 套件，包括电子邮件客户端……

In client/server computing, processing is distributed between two machines——the client machine and the server machine. 在客户机/服务器计算中，处理工作分布在两台机器上进行——客户机和服务器。

用作计算机专业词汇时，client 可以表示"客户程序"（软件领域），也可以表示"客户机"（硬件领域）。第一句话中的 client 显然指程序，第二句中的 client 与 server 相对，显然指硬件。

另外需要说明的是：有的计算机术语不加翻译可以直接借用到汉语中，并已经为IT业所惯用，如句中出现的 Office，还有 Java、IBM 等。

例句3：

Because transistors use much less power and have a much longer life…因为晶体管的功率更小，寿命更长……

power 有"权力、力量、功率、动力、政权"等意思，根据上下文，此处介绍晶体管与电子管的区别，结合对当前计算机业的了解，生产的硬件大小有越来越小的趋势，而运算能力、使用寿命、功率等都有所提高。因此，此处的 power 翻译成"功率"更合理。

3.2 词性、句子成分、句子类型转换

英语和汉语之间有许多不同，其中对翻译影响较大的就是英语多用形合，汉语多用意合。所谓形合，就是在形式上加以配合，使用复句或从句，用一种语法形式来表达它们之间的句法关系，句中逻辑关系很严密，通常使用关联词语来实现。所谓意合，就是在意思上加以配合，通过上下文的逻辑关系来约束各个分句之间的关系。

英译汉时，如果拘泥于原文单词或句子的语法特征，势必产生拗口的句子甚至难以理解的句子。因此，在忠实于原文的前提下，要敢于突破语法框框，对词性、句子成分、句子类型等进行变通转换，以使译文自然流畅，符合汉语习惯。

例句4：

The new electronic computer is chiefly characterized by its simplicity of structure. 这种新型计算机的主要特点是结构简单。

本句涉及副词转换为形容词，即副词 chiefly 译文中变成了"主要的"。当英语动词 characterize 在翻译时转换为汉语名词时，原来相关的状语副词往往可以译为形容词。

例句 5：

A dominant factor in the growth of MS in question throughout the years has been its success in maintaining technical superiority in product development. 微软公司这些年来发展壮大的主要因素是其一直成功地保持了在产品开发方面的技术优势。

本句涉及名词转换为副词，即名词 success 在译文中变成了副词"成功地"。

例句 6：

You should use your E-mail account with the expectation that some of your mail will be read from time to time. 使用电子邮件账号时，你应该想到：你的一些邮件会被人不时地阅读。

本句涉及名词转换为动词，即名词 expectation 在译文中变成了动词"想到"。此外，原文的句子成分在译文中也发生了变化：主句的谓语动词 use 及其宾语 your E-mail account 转为时间状语；主句作状语的介词短语 with the expectation 转为主句的谓语动词；同位语从句转为宾语从句。这句话如果按照原文的单词词性和句子成分翻译，会非常拗口：你应该带着你的一些邮件会被人不时阅读的料想使用你的电子邮件账号。

例句 7：

The properties of this computer should be made full use of. 应该充分利用这台计算机的性能。

本句的主语转换为宾语。

例句 8：

The first inventor of this precision instrument seems to have been Charles Babbage in 1847—the same man who is regarded as the father of the computer. 计算机之父查尔斯·巴贝奇在 1847 年首先发明了这种精密仪器。

本句是表语转换为主语。

3.3 增词与减词

在英译汉时，译文增添或者省略一些词，是为了更好地表达原文的意思，便于读者理解原文。但是，补充只能在原文的基础上和作者的意思内进行，不能离题发挥。省略的词必须是译文中确实不需要表达的，不能损及原文的意思。

例句 9：

Computers have opened up a new era in manufacturing through the techniques of automation, and they have enhanced modern communication systems. 通过自动化技术，计算机开辟了制造业的新纪元，也增强了现代通信系统的性能。

本句通过增词将原文的意思表达完整。

例句 10：

It would be easy to assume that computers will all be able to work together once they have broadband connection. 人们很容易认为：计算机一旦有了宽带连接，就能全部具备协同工作的能力。

这句话中的 It 是形式主语，真正主语是 that 引导的从句，翻译时增加了泛指的"人们"来作主语。

例句 11：

In most wire lines, two conductors, a go and a return, are used for signal transmission. 在大多数有线线路中,用两根导线:一根去线和一根回线来传输信号。

本句中的 go 和 return 是抽象名词,当它们被赋予某种具体的含义时,需要补充适当的、具体的名词来表达含义。

例句 12：

In the implementation phase, you create the actual programs. 在实现阶段,建立真正的程序。

本句中的人称代词 you 是泛指,常常不译。

例句 13：

A database system is divided into several modules to achieve the overall functionality. 数据库系统通过分成几个模块来完成总体功能。

这里的不定冠词 a 表示一类,而非数量一,不译出来,译文更简洁,更顺畅。

例句 14：

In the design phase, the systems are determined, and the design of the files and/or the databases is completed. 设计阶段需要确定系统,完成文件和/或数据库的设计。

英语求形合,并列关系通常加上 and,汉语求意合,句中的逻辑关系往往不言自明。因此,在本句中,the systems are determined 和 the design of the files and/or the databases is completed 两个并列的分句,翻译时去掉了 and,使译文更符合汉语的习惯。

Unit Four Information Management

Section A Information Storage

Ⅰ. The Importance of Information

Information is increasingly important in our daily lives. We have become information dependent in the 21st century, living in an on-command, on-demand world, which means, we need information when and where it is required. We access the Internet every day to perform searches, participate in social networking, send and receive E-mails, share pictures and videos, and use scores of other applications. Equipped with a growing number of content-generating devices, more information is created by individuals than by organizations. Information created by individuals gains value when shared with others. When created, information **resides** locally on devices, such as cell phones, smartphones, **tablets**, cameras, and **laptops**. To be shared, this information needs to be uploaded to central data **repository** (data centers) via networks. Although the majority of information is created by individuals, it is stored and managed by a relatively small number of organizations.

The importance, dependency, and volume of information for the business world also continue to grow at astounding rates. Businesses depend on fast and reliable access to information critical

reside
/rɪˈzaɪd/
vi. 属于;驻留
tablet
/ˈtæblɪt/
n. 平板电脑
laptop
/ˈlæptɒp/
n. 笔记本电脑
repository
/rɪˈpɒzət(ə)ri/
n. 仓库

to their success. Examples of business processes or systems that rely on digital information include airline reservations, telecommunications billing, internet commerce, electronic banking, credit card transaction processing, capital/stock trading, health care **claims** processing, life science research and so on. The increasing dependence of businesses on information has amplified the challenges in storing, protecting, and managing data. Legal, **regulatory**, and **contractual obligations** regarding the availability and protection of data further add to these challenges.

II. What is Data

Data is a collection of raw facts form which conclusions may be drawn. Handwritten letters, a printed book, a family photograph, printed and duly signed copies of mortgage papers, a bank's ledgers, and an airline ticket are examples that contain data.

Before the advent of computers, the methods adopted for data creation and sharing were limited to fewer forms, such as paper and film. Today, the same data can be converted into more convenient forms, such as an E-mail message, an E-book, a digital image, or a digital movie. This data can be generated using a computer and stored as strings of **binary** numbers(0s and 1s). Data in this form is called **digital** data and is accessible by the user only after a computer processes it.

Data can be classified as structured or unstructured based on how it is stored and managed. Structured data is organized in rows and columns in a rigidly defined format so that applications can retrieve and process it efficiently. Structured data is typically stored using a **database management system(DBMS)**.

Data is unstructured if its elements cannot be stored in rows and columns, which makes it difficult to **query** and **retrieve** by applications. For example, customer contacts that stored in various forms such as sticky notes, E-mail messages, business cards, or even digital format files, such as .doc, .txt, and .pdf. Due to its unstructured nature, it is difficult to retrieve this data using a traditional customer relationship management application. A vast majority of new data being created today is unstructured. The industry is challenged with new architectures, technologies,

---

claim
/kleɪm/
n. 索赔
regulatory
/ˈreɡjʊleɪtri/
n. 法规
contractual
/kənˈtræktʃʊəl/
adj. 合同的
obligation
/ˌɒblɪˈɡeɪʃn/
n. 合约；责任

binary
/ˈbaɪnəri/
n. 二进制
digital
/ˈdɪdʒɪtl/
adj. 数字的

DBMS
数据库管理系统
query
/ˈkwɪəri/
v. 查询
retrieve
/rɪˈtriːv/
v. 检索

techniques, and skills to store, manage, analyze and derive value from unstructured data form numerous sources.

III. Evolved of Storage Architecture

Data created by individuals or businesses must be stored so that it iseasily accessible for further processing. In a computing environment, devices designed for storing data are **termed** storage devices or simply storage. The type of storage used varies based on the type of data and the rate at which it is created and used. Devices, such as a media card in a cell phone or digital camera, DVDs, CD-ROMs, and disk drives in personal computers are examples of storage devices.

Businesses have several options available for storing data, including internal hard disks, external disk arrays, and tapes.

Historically, organizations had centralized computers (**mainframes**) and information storage devices (**tape reels** and **disk packs**) in their data center. The evolution of open systems, their **affordability**, and ease of **deployment** made it possible for business departments to have their own servers and storage. In earlier implementations of open systems, the storage was typically internal to the server. These storage devices could not be shared with any other servers. This approach is referred to server-centric storage architecture. In this architecture, each server has a limited number of storage devices, and any administrative tasks, such as **maintenance** of the server or increasing storage capacity, might result in unavailability of information. The proliferation of departmental servers in an enterprise resulted in unprotected, unmanaged fragmented islands of information and increased capital and operating expenses.

To overcome these challenges, storage evolved from server-centric to information-centric architecture (see Figure 4A-1). In this architecture, storage devices are managed centrally and independent of servers. These centrally-managed storage devices are shared with multiple servers. When a new server is deployed in the environment, storage is assigned from the same shared storage devices, to that server. The capacity of shared storage can be increased dynamically by adding more storage devices without

storage
/ˈstɔːrɪdʒ/
n. 存储

term
/tɜːm/
vt. 把……称为

mainframe
/ˈmeɪnfreɪm/
n. 主机；大型机
tape reel
磁带盘
disk pack
磁盘组
affordability
/əˌfɔːdəˈbɪləti/
n. 可购性；成本合理性
deployment
/dɪˈplɔɪmənt/
n. 部署
maintenance
/ˈmeɪntənəns/
n. 维护；维修

impacting information availability. In this architecture, information management is easier and cost-effective.

Storage technology and architecture continue to evolve, which enables organizations to **consolidate**, protect, optimize, and **leverage** their data to achieve the highest return on information assets.

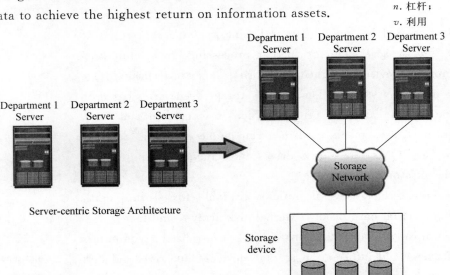

Figure 4A-1 Evolved of Storage Architecture

Ⅳ. Storage Networking Technologies

- **SAN(Storage Area Network)**

SAN is a high-speed, dedicated network of servers and shared storage devices. It enables sharing of storage resources across multiple servers at block level. Common SAN deployments are: **Fibre** Channel(FC) SAN(see Figure 4A-2) and IP SAN. The former uses FC protocol for communication and the latter uses IP-based protocols for communication.

FC has high-speed network technology, the latest FC implementation supports speed up to 16 **Gb**/s. Also it is highly **scalable**, theoretically, accommodate approximately 15 million devices.

A technique which transports block-level data over IP network is called IP SAN. IP is being positioned as a storage networking option because: Existing network infrastructure can be leveraged; Reduce cost compared to investing in new FC SAN hardware and software; Many long-distance disaster recovery solutions already leverage IP-based network; Many **robust** and mature security options are available for IP network.

robust
/rəʊˈbʌst/
adj. 强健;鲁棒

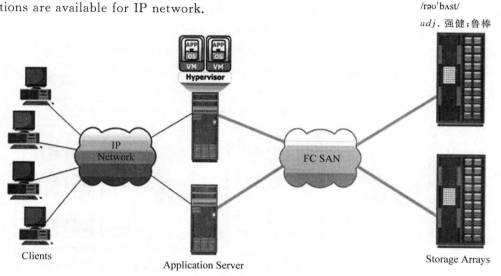

Figure 4A-2　FC SAN

- **NAS(Network Attached Storage)**

NAS(See Figure 4A-3) is an IP-based, dedicated, high-performance file sharing and storage device. It enables NAS clients to share files over IP network, uses specialized operating system that is optimized for file I/O, and enables both UNIX and Windows users to share data **seamlessly**.

NAS
网络连接存储
seamlessly
/ˈsiːmlɪsli/
adv. 无缝地

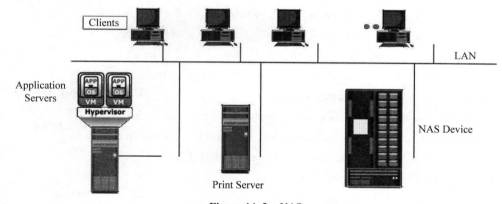

Figure 4A-3　NAS

- **Object-based Storage**

More than 90% of the data being generated is unstructured. Traditional solutions are inefficient to handle the growth. These challenges demanded a smarter approach to manage unstructured data based on its content. Object-based storage is a way to store file data in the form of objects on **flat** address space based on its content and attributes rather than the name and location. Figure 4A-4 displays the key components of Object-based Storage device.

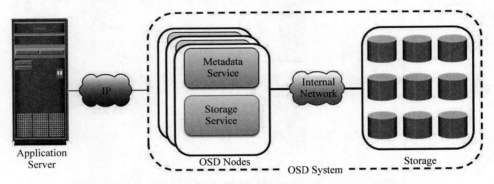

Figure 4A-4 The Key Components of Object-based Storage Device

Ⅴ. Challenge of Storage

Data science enables organizations to derive business value from big data. Several industries and markets currently looking to employ data science techniques include medical and scientific research, healthcare, public administration, fraud detection, social media, banks, insurance companies, and other digital information-based entities that benefit from the analytics of big data. The storage architecture required for big data should be simple, efficient, and inexpensive to manage, yet provide access to multiple platforms and data sources **simultaneously**.

Organizations maintain data centers to provide centralized data-processing capabilities across the enterprise. Data center is a facility that contains storage, compute, network, and other IT resources to provide centralized data-processing capabilities.

Virtualization and cloud computing have dramatically changed the way data center infrastructure resources are **provisioned** and managed. Organizations are rapidly deploying virtualization on various elements of data centers to optimize their utilization. Further, continuous cost pressure on IT and on-demand data processing requirements have resulted in the adoption of cloud computing. Cloud infrastructure is usually built upon virtualized data centers, which provide resource pooling and rapid provisioning of resources.

Exercises

I. Fill in the blanks with the information given in the text.

1. Although the majority of information is created by individuals, it is _____ and _____ by a relatively small number of organizations.

2. _____ is a collection of raw facts form which conclusions may be drawn.

3. This data can be generated using a computer and stored as strings of _____ numbers(0s and 1s).

4. Data can be classified as structured or _____ based on how it is stored and managed.

5. Structured data is typically stored using a _____.

6. In _____ architecture, storage devices are managed centrally and independent of servers.

7. _____ is a facility that contains storage, compute, network, and other IT resources to provide centralized data-processing capabilities.

8. Cloud infrastructure is usually built upon _____ data centers, which provide resource pooling and rapid provisioning of resources.

II. Translate the following terms or phrases from English into Chinese.

data center	binary
digital	data science
DBMS	mainframe
tape reel	disk pack
SAN	Fibre Channel(FC)
scalable	robust
NAS	object-based storage

Ⅲ. Translate the following passage from English into Chinese.

Big Data

Big data is a new and evolving concept, which refers to data sets whose sizes are beyond the capability of commonly used software tools to capture, store, manage, and process within acceptable time limits. It includes both structured and unstructured data generated by a variety of sources, including business application transactions, web pages, videos, images, E-mails, social media, and so on. These data sets typically require real-time capture or updates for analysis, predictive modeling, and decision making.

Traditional IT infrastructure and data processing tools and methodologies are inadequate to handle the volume, variety, dynamism, and complexity of big data. Analyzing big data in real time requires new techniques, architectures, and tools that provide high performance, massively parallel processing(MPP) data platforms, and advanced analytics on the data sets.

Section B Data Mining

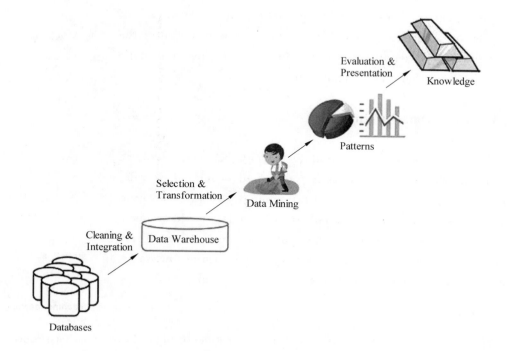

Ⅰ. Introduction

A rapidly expanding subject that is closely associated with database technology is **data mining** (sometimes called knowledge discovery in database, KDD), which consists of techniques for discovering patterns in collections of data. Data mining has become an important tool in numerous areas including marketing, inventory management, quality control, loan risk management, fraud detection, and investment analysis. Data mining software is one of a number of analytical tools for analyzing data. It allows users to analyze data from many different dimensions or angles, categorize it, and summarize the relationships identified. Technically, data mining is the process of finding correlations or patterns among dozens of fields in large relational databases.

Data mining activities differ from traditional database **interrogation** in that data mining seeks to identify previously unknown patterns as opposed to traditional database inquiries that

data mining
数据挖掘

interrogation
/ɪnˌterəˈgeɪʃn/
n. 询问

merely ask for the retrieval of stored facts. Moreover, data mining is practiced on static data collections, called **data warehouses**, rather than "online" operational databases that are subject to frequent updates. These warehouses are often "snapshots" of databases or collections of databases. They are used in **lieu** of the actual operational databases because finding patterns in a static system is easier than in a dynamic one.

Ⅱ. Forms of Data Mining

While large-scale information technology has been evolving separate transaction and analytical systems, data mining provides the link between the two. Data mining software analyzes relationships and patterns in stored transaction data based on open-ended user queries. Several types of analytical software are available: **statistics**, **machine learning**, and **neural networks**. Generally, any of five types of relationships are sought:

- **Classes**

Stored data is used to locate data in predetermined groups. For example, a restaurant chain could mine customer purchase data to determine when customers visit and what they typically order. This information could be used to increase traffic by having daily specials.

- **Cluster**

Cluster analysis seeks to discover classes. Note that this differs from class description, which seeks to discover properties of members within classes that are already identified. More precisely, cluster analysis tries to find properties of data items that lead to the discovery of groupings.

- **Association**

Association analysis involves looking for links between data groups. It is association analysis that might reveal that customers who buy potato chips also buy beer and soda or that people who shop during the traditional weekday work hours also draw retirement benefits.

---

data warehouse
数据仓库
lieu
/ljuː/
n. 代替；场所

statistics
/stəˈtɪstɪks/
n. 统计学
machine learning
机器学习
neural network
神经网络

cluster analysis
聚类分析

association analysis
关联分析

- **Outlier**

Outlier analysis is another form of data mining. It tries to identify data entries that do not comply with the norm. Outlier analysis can be used to identify errors in data collections, to identify credit card theft by deviation detecting sudden **deviations** from a customer's normal purchase patterns, and perhaps to identify potential terrorists by recognizing unusual behavior.

- **Sequential patterns**

Finally, the form of data mining called **sequential pattern analysis** tries to identify patterns of behavior over time. For example, sequential pattern analysis might reveal trends in economic systems such as equity markets or in environmental systems such as climate conditions.

Ⅲ. Data Mining Applications

Nowadays, data mining is widely used in most fields. For example:
- Social network applications
- Scalable data preprocessing and cleaning techniques
- Data mining systems in finance, sciences, retail, e-commerce
- Emerging applications of large-scale data mining
- **Empirical** study of data mining algorithms
- Parallel data mining applications
- DNA Sequencing, **bioinformatics**, **genomics**, and **biometrics**
- E-commerce and Web services
- Medical informatics
- Disaster prediction
- Financial market analysis
- Intelligent system
- Application of data mining in education

Successful data mining encompasses much more than the identification of patterns within a collection of data. Intelligent judgment must be applied to determine whether those patterns are significant or merely **coincidences**. The fact that a particular convenience store has sold a high number of winning lottery tickets should probably not be considered significant to someone planning to buy a lottery ticket, but the discovery that customers who buy snack food also tend to buy frozen dinners might constitute meaningful information to a grocery store manager. Likewise, data mining encompasses a vast number of **ethical** issues involving the rights of individuals represented in the data warehouse, the accuracy and use of the conclusions drawn, and even the appropriateness of data mining in the first place.

Although data mining is a relatively new term, the technology is not. Companies have used powerful computers to sift through volumes of supermarket scanner data and analyze market research reports for years. However, continuous innovations in computer processing power, disk storage, and statistical software are dramatically increasing the accuracy of analysis while driving down the cost.

ethical
/'eθɪkl/
n. 道德的；民族的

Exercises

Ⅰ. Fill in the blanks with the information given in the text.

1. A rapidly expanding subject that is closely associated with database technology is _____, which consists of techniques for discovering _____ in collections of data.

2. Data mining is the process of finding _____ or patterns among dozens of fields in large relational databases.

3. Several types of analytical software in data mining are available: statistical, machine learning, and _____.

4. Association analysis involves looking for _____ between data groups.

5. _____ analysis tries to identify data entries that do not comply with the norm.

6. Data mining encompasses a vast number of _____ issues involving the rights of individuals represented in the data warehouse.

Ⅱ. Translate the following terms or phrases from English into Chinese.

data mining   knowledge discovery in data(KDD)
data warehouses   machine learning
neural networks   cluster analysis
association analysis   outlier analysis
sequential pattern analysis

Section C Data Center

Ⅰ. What is Data Center

Organizations maintain data centers to provide **centralized** data-processing capabilities across the enterprise. Data centers **house** and manage large amounts of data. The data center infrastructure includes hardware components, such as computers, storage systems, networks devices, and power backups; and software components, such as applications, operations systems, and management software. It also includes environmental controls, such as air conditioning, fire **suppression**, and **ventilation**.

Large organizations often maintain more than one data center to distribute data processing workloads and provide backup if a disaster occurs.

Ⅱ. The Core Elements of Data Center

Five core elements are essential for the functionality of a data center:
- Application: A computer program that provides the logic for computing operations.
- Database management system(DBMS): Provides a structured

way to store data in logically organized tables that are interrelated.
- Host or compute: A computing **platform**(hardware, **firmware** and software) that runs applications and databases.
- Network: A data path that facilitates communication among various networked devices.
- Storage: A device that stores data persistently for subsequent use.

These core elements are typically viewed and managed as separate entities, but all the elements must work together to address data-processing requirements.

Figure 4C-1 shows an example of an online order transaction system that involves the five core elements of a data center and illustrates their functionality in a business process.

platform
/'plætfɔːm/
n. 平台
firmware
/'fɜːmweə(r)/
n. 固件

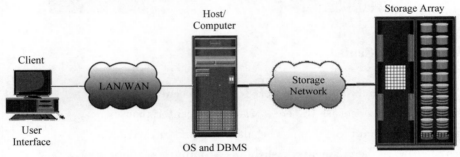

Figure 4C-1　An Online Order Transaction System

A customer places an order through a client machine connected over a LAN/WAN to a host running an order-processing application. The client accesses the DBMS on the host through the application to provide order-related information, such as the customer name, address, payment method, products ordered, and quantity ordered.

The DBMS uses the host operating system to write this data to the physical disks in the storage array. The storage networks provide the communication link between the host and the **storage array** and transports the request to read or write data between them. The storage array, after receiving the read or write request from the host, performs the necessary operations to store the data on physical disks.

storage array
存储阵列

Ⅲ. Characteristics of Data Center

Uninterrupted operation of data centers is critical to the survival and success of a business. The characteristics are applicable to all elements of the data center infrastructure.

- **Availability**

A data center should ensure the availability of information when required. **Unavailability** of information could cost millions of dollars per hour to businesses, such as financial services, telecommunications, and e-commerce.

- **Security**

Data centers must establish policies, procedures, and core element **integration** to prevent **unauthorized** access to information.

- **Scalability**

Business growth often requires deploying more servers, new applications, and additional databases. Data center resources should scale based on requirements, without interrupting business operations.

- **Performance**

All the elements of the data center should provide optimal performance based on the required service levels.

- **Data integrity**

Data integrity refers to mechanisms, such as error correction codes or **parity** bits, which ensure that data is stored and retrieved exactly as it was received.

- **Capacity**

Data center operations require adequate resources to store and process large amounts of data efficiently. When capacity requirements increase, the data center must provide additional capacity without interrupting availability or with minimal disruption. Capacity may be managed by reallocation the existing

uninterrupted
/ˌʌnɪntəˈrʌptɪd/
adj. 不间断的

unavailability
/ˌʌnəˈveɪləbɪlɪti/
n. 无效

integration
/ˌɪntɪˈgreɪʃn/
n. 集成；融合
unauthorized
/ʌnˈɔːθəraɪzd/
adj. 未经授权的

integrity
/ɪnˈtegrəti/
n. 完整性
parity
/ˈpærəti/
n. 奇偶
capacity
/kəˈpæsəti/
n. 容量；能力

resources or by adding new resources.

- **Manageability**

A data center should provide easy and integrated management of all its elements. Manageability can be achieved through automation and reduction of human **intervention** in common tasks.

intervention
/ˌɪntəˈvenʃn/
n. 介入；干预

Exercises

Ⅰ. Fill in the blanks with the information given in the text.

1. The hardware components of data center infrastructure include computers, _____, networks devices, and _____.

2. _____ operation of data centers is critical to the survival and success of a business.

3. Data centers must establish policies, procedures, and core element integration to prevent unauthorized _____ to information.

Ⅱ. Translate the following terms or phrases from English into Chinese.

data center	power backup
workload	backup
DBMS	storage array
unauthorized access	parity bit

Unit Five Networking

Section A Networking

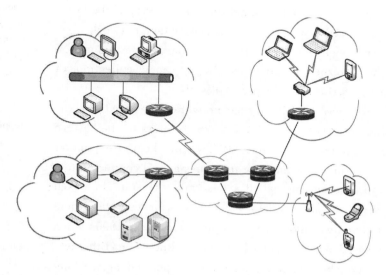

The need to share information and resources among different computers has led to linked computer systems, called **networks**, in which computers are connected so that data can be transferred from machine to machine.

network
/ˈnetwɜːk/
n. 网络

Ⅰ. Network Classifications

A computer network is often classified as being either a local area network(**LAN**), a **metropolitan** area network(**MAN**), or a wide area network(**WAN**). In recent years, with the development of wireless communication technology, a new concept, personal area network(**PAN**) is put forward. A LAN normally consists of a collection of computers in a single building or building complex. For example, the computers on a university campus or those in a manufacturing plant might be connected by a LAN. A MAN is a network of intermediate size, such as one spanning a local community. A WAN links machines over a greater distance—perhaps in neighboring cities or on opposite sides of the world. A

LAN
局域网
metropolitan
/ˌmetrəˈpɒlɪtən/
n. 大都市的
MAN
城域网
WAN
广域网
PAN
个域网

PAN is a personal information network, as **cable radio** or **infrared** instead of the traditional cable, to realize the intelligent internet of personal information terminal.

Another means of classifying networks is based on whether the network's internal operation is based on designs that are in the public domain or on innovations owned and controlled by a particular entity such as an individual or a corporation. A network of the former type is called an **open network**; a network of the latter type is called a closed, or sometimes a **proprietary** network. The Internet is an open system. In particular, communication throughout the Internet is governed by an open collection of standards known as the TCP/IP protocol suite.

Still another way of classifying networks is based on the **topology** of the network (Figure 5A-1), which refers to the pattern in which the machines are connected. Two of the more popular topologies are the bus, in which the machines are all connected to a common communication line called a **bus**, and the star, in which one machine serves as a central focal point to which all the others are connected. The bus topology was popularized in the 1990s when it was implemented under a set of standards known as **Ethernet**, and Ethernet networks remain one of the most popular networking systems in use today. The star topology has roots as far back as the 1970s. It evolved from the paradigm of a large central computer serving many users. Another topologies are the ring topology, tree topology, **mesh** topology and **hybrid** topology.

cable radio
无线电
infrared
/ˌɪnfrəˈred/
n. 红外线

open network
开放网络
proprietary
/prəˈpraɪətri/
adj. 专有的

topology
/təˈpɒlədʒɪ/
n. 拓扑学

bus
/bʌs/
n. 总线

Ethernet
/ˈiːθənet/
n. 以太网
mesh
/meʃ/
n. 网状;网格
hybrid
/ˈhaɪbrɪd/
adj. 混合的

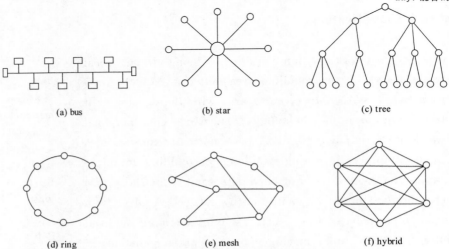

(a) bus   (b) star   (c) tree

(d) ring   (e) mesh   (f) hybrid

Figure 5A-1   The Topology of the Network

II. Combining Networks

Sometimes it is necessary to connect existing networks to form an extended communication system. This can be done by connecting the networks to form a larger version of the same type of network. In the case of bus networks based on the Ethernet protocols, it is often possible to connect the buses to form a single long bus. This is done by means of different devices known as **repeaters**, **bridges**, and **switches**, the distinctions of which are **subtle** yet **informative**. The simplest of these is the repeater, which is little more than a device that simply passes signals back and forth between the two original buses (usually with some form of **amplification**) without considering the meaning of the signals.

A bridge is similar to, but more complex than, a repeater. Like a repeater, it connects two buses, but it does not necessarily pass all messages across the connection. Instead, it looks at the destination address that accompanies each **message** and **forwards** a message across the connection only when that message is destined for a computer on the other side. Thus, two machines residing on the same side of a bridge can exchange messages without interfering with communication taking place on the other side. A bridge produces a more efficient system than that produced by a repeater.

A switch is essentially a bridge with multiple connections, allowing it to connect several buses rather than just two. Thus, a switch produces a network consisting of several buses extending from the switch as spokes on a wheel. As in the case of a bridge, a switch considers the destination addresses of all messages and forwards only those messages destined for other spokes. Moreover, each message that is forwarded is relayed only into the appropriate spoke, thus minimizing the traffic in each spoke.

Sometimes, however, the networks to be connected have incompatible characteristics. For instance, the characteristics of a **WiFi**① network are not readily compatible with an Ethernet network. In these cases the networks must be connected in a manner that builds a network of networks, known as an internet,

---

① WiFi: 基于 IEEE 802.11b 标准的无线局域网。

repeater
/rɪˈpiːtə(r)/
n. 中继器
bridge
/brɪdʒ/
n. 网桥
switch
/swɪtʃ/
n. 交换机
subtle
/ˈsʌtl/
adj. 微妙的
informative
/ɪnˈfɔːmətɪv/
adj. 信息量大的
amplification
/ˌæmplɪfɪˈkeɪʃn/
n. 放大
message
/ˈmesɪdʒ/
n. 报文
forward
/ˈfɔːwəd/
vt. 转发

in which the original networks maintain their individuality and continue to function as **autonomous** networks. The connection between networks to form an internet is handled by devices known as **routers**, which are special purpose computers used for forwarding messages.

As an example, Figure 5A-2 depicts two WiFi star networks and an Ethernet bus network connected by routers. When a machine in one of the WiFi networks wants to send a message to a machine in the Ethernet network, it first sends the message to the **AP**[①] in its network. From there, the AP sends the message to its associated router, and this router forwards the message to the router at the Ethernet. There the message is given to a machine on the bus, and that machine then forwards the message to its final destination in the Ethernet.

autonomous
/ɔː'tɒnəməs/
adj. 自主的
router
/'ruːtə/
n. 路由器

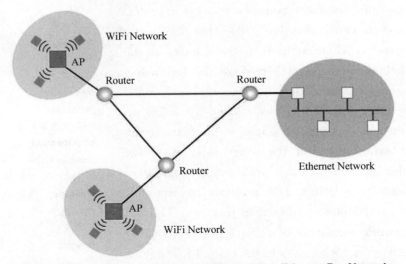

Figure 5A-2　Routers Connecting Two WiFi and an Ethernet Bus Network

The point at which one network is linked to an internet is often called a **gateway** because it serves as a passageway between the network and the outside world. Gateways can be found in a variety of forms, and thus the term is used rather loosely. In many cases a network's gateway is merely the router through which it communicates with the rest of the internet. In other cases the term gateway may be used to refer to more than just a router. For

gateway
/'geɪtweɪ/
n. 网关

---

① AP：Access Point，即（无线）访问接入点，是组建小型无线局域网最常用的设备。

example, in most residential WiFi networks that are connected to the Internet, the term gateway refers collectively to both the network's AP and the router connected to the AP because these two devices are normally packaged in a single unit.

III. Methods of Process Communication

The various activities(or processes) executing on the different computers within a network(or even executing on the same machine via time-sharing/multitasking) must often communicate with each other to coordinate their actions and to perform their designated tasks. Such communication between processes is called **interprocess** communication.

A popular convention used for interprocess communication is the client/server model. This model defines the basic roles played by the processes as either a client, which makes requests of other processes, or a server, which satisfies the requests made by clients. Today the client/server model is used extensively in network applications.

However, the **client/server(C/S)** model is not the only means of interprocess communication. Another model is the **peer-to-peer (P2P)** model. Whereas the client/server model involves one process (the server) providing a service to numerous others(clients), the peer-to-peer model involves processes that provide service to and receive service from each other(Figure 5A-3). Moreover, whereas a server must execute continuously so that it is prepared to serve its clients at any time, the peer-to-peer model usually involves processes that execute on a temporary basis. For example, applications of the peer-to-peer model include instant messaging in which people carry on a written conversation over the Internet as well as situations in which people play competitive interactive games.

The peer-to-peer model is also a popular means of distributing files such as music recordings and motion pictures via the Internet. In this case, one peer may receive a file from another and then provide that file to other peers. The collection of peers participating in such a distribution is sometimes called a **swarm**. The swarm approach to file distribution is in contrast to earlier approaches that

interprocess
/ˈɪntəˌprɒses/
n. 进程间

client/server
客户/服务器
peer-to-peer
对等网;点对点

swarm
/swɔːm/
n. 聚合;群

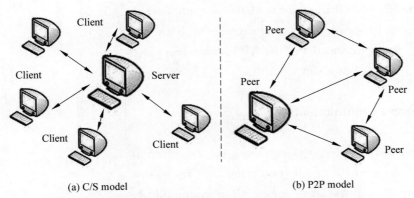

Figure 5A-3 Two Popular Models of Interprocess Communication

applied the client/server model by establishing a central distribution center(the server) from which clients downloaded files.

Exercises

I . Fill in the blanks with the information given in the text.

1. The need to share information and _____ among different computers has led to linked computer systems, called _____, in which computers are connected so that data can be transferred from machine to machine.

2. A computer network is often classified as being either a _____, a MAN, a WAN, or a _____.

3. The Internet is an _____ system, which is governed by an open collection of standards known as the _____ protocol suite.

4. Based on the topology of the network, two of the more popular topologies are _____ , and _____ .

5. A _____ is essentially a bridge with multiple connections, allowing it to connect several buses rather than just two.

6. The connection between networks to form an internet is handled by devices known as _____ .

7. The point at which one network is linked to an internet is often called a _____ because it serves as a passageway between the network and the outside world.

8. In most residential WiFi networks that are connected to the Internet, the term gateway refers collectively to both the network's _____ and the _____ connected to the AP because these two devices are normally packaged in a single unit.

9. The popular convention used for interprocess communication is the _____ model and _____ model.

II. Translate the following terms or phrases from English into Chinese.

LAN	MAN
WAN	PAN
cable radio	open network
proprietary network	bus topology
Ethernet network	ring topology
tree topology	mesh topology
hybrid topology	repeater
bridge	switch
router	WiFi
AP	gateway
client/server(C/S)	peer-to-peer(P2P)

III. Translate the following passage from English into Chinese.

The Generations of Wireless Telephones

In the past decade mobile phone technology has advanced from simple, single purpose, portable devices to complex, multifunction hand-held computers. The first generation wireless telephone network transmitted analog voice signals through the air, much like traditional telephones but without the copper wire running through the wall. In retrospect, we call these early phone systems "1G", which using FDMA. The second generation mainly using TDMA, can provide voice and low-speed digital services such as text messaging. Third generation(3G) phone network which using CDMA technology is characterized by user peak rate of 2Mb/s to reach tens of Mb/s, allowing for mobile video calls and other multimedia data services. 4G OFDMA technology as the core, the user peak rate of up to 100Mb/s~1Gb/s, can support a variety of mobile broadband data services. 5G key competencies richer than previous generations of mobile communications, user experience, speed, density of connections, end to end delay, the peak rate and mobility and so will be the 5G key performance indicators.

Section B Distributed System

Ⅰ. Introduction

With the success of networking technology, interaction between computers via networks has become common and **multifaceted**. Many modern software systems, such as global information retrieval systems, company-wide accounting and inventory systems, computer games, and even the software that controls a network's infrastructure itself are designed as **distributed** systems, meaning that they consist of software units that execute as processes on different computers.

A distributed system (see Figure 5B-1) consists of a collection of autonomous computers, connected through a network and distribution **middleware**, which enables computers to coordinate their activities and to share the resources of the system, so that users perceive the system as a single, integrated computing facility.

The certain common characteristics can be used to assess distributed systems:
- Resource Sharing
- Openness
- **Concurrency**
- Scalability
- Fault **Tolerance**
- **Transparency**

Middleware is a class of software technologies designed to help

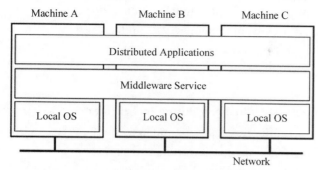

Figure 5B-1 General Structure of a Distributed System

manage the complexity and **heterogeneity** **inherent** in distributed systems. It is defined as a layer of software above the operating system but below the application program that provides a common programming abstraction across a distributed system.

II. Application of Distributed System

Several types of distributed computing systems are now common.

- **Cluster Computing**

Cluster computing(Figure 5B-2) is a type of parallel or distributed processing system, which consists of a collection of interconnected stand-alone computers cooperatively working together as a single, integrated computing resource. The cost of these individual machines, plus the high-speed network to connect them, can be less than a higher-priced supercomputer, but with higher reliability and lower maintenance costs. Such distributed systems has the key operational benefits of high performance, expandability and scalability, high **throughput** and high availability.

Figure 5B-2 Cluster Computing

Cluster computing is mainly used in grand challenging applications such as weather forecasting, quantum chemistry, molecular biology modeling, data-mining web servers, etc.

- **Grid Computing**

Grid computing is a form of distributed computing in which an organization uses its existing computers(desktop and/or cluster nodes) to handle its own long-running computational tasks(see Figure 5B-3). Grid computing aims to "enable resource sharing and coordinated problem solving in dynamic, multi-institutional virtual organizations". Examples include University of Wisconsin's Condor system[①], or Berkeley's Open Infrastructure for Network Computing(BOINC)[②]. Both of these systems are often installed on computers that are used for other purposes, such as PCs at work or at home, that can then **volunteer** computing power to the grid. Enabled by the growing connectivity of the Internet, this type of **voluntary**, distributed grid computing has enabled millions of home PCs to work on enormously complex mathematical and scientific problems.

grid
/grɪd/
n. 网格

volunteer
/ˌvɒlənˈtɪə(r)/
v. 自愿去做/
n. 自(志)愿者
voluntary
/ˈvɒləntri/
n. 自愿；
adj. 自愿的

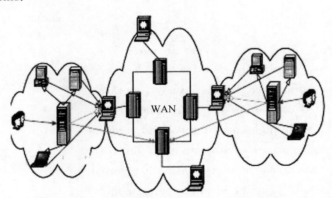

Figure 5B-3　A Distributed Grid Task Scheduling System

- **Cloud Computing**

Cloud computing(Figure 5B-4) is a large-scale distributed

---

① Wisconsin's Condor system：威斯康星大学研发的 Condor 系统，是一个开源的高吞吐量计算软件框架，用于计算密集型任务的粗粒度分布式并行计算。
② BOINC：伯克利开放式网络计算，由美国加利福尼亚大学伯克利分校空间科学实验室研发的用于志愿者和网格计算的一个开源的中间件系统。

computing paradigm that is driven by economies of scale, in which a pool of abstracted, virtualized, **dynamically**-scalable, managed computing power, storage, platforms, and services are delivered on demand to external customers over the Internet. Cloud computing is hinting at a future in which we won't compute on local computers, but on centralized facilities operated by third-party compute and storage utilities.

dynamically
/daɪˈnæmɪkli/
adv. 动态地

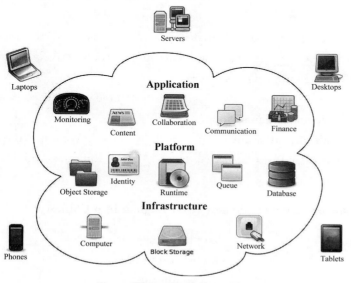

Figure 5B-4　Cloud Computing

Services such as Amazon's Elastic Compute Cloud[①] allow clients to rent virtual computers by the hour, without concern for where the computer hardware is actually located. Google Docs[②] and Google Apps[③] allow users to collaborate on information or build Web services without needing to know how many computers are working on the problem or where the relevant data are stored. Cloud computing services provide reasonable guarantees of reliability and scalability, but also raise concerns about privacy and security in a world where we may no longer know who owns and operates the computers that we use.

---

① Amazon's Elastic Compute Cloud：亚马逊弹性云计算（简称 Amazon EC2），是亚马逊公司提供的 Web 服务，利用其全球性的数据中心网络，为客户提供虚拟主机服务。
② Google Docs：谷歌办公套件，类似于微软 Office 的一套在线办公软件，可以处理和搜索文档、表格、幻灯片，并可以通过网络和他人分享。
③ Google Apps：Google 提供的一项"软件即服务"产品，用于企业消息传输、协作和安全。

Exercises

I. Fill in the blanks with the information given in the text.

1. A distributed system consists of a collection of autonomous computers, connected through a network and distribution _____.

2. Middleware is a class of _____ technologies designed to help manage the complexity and heterogeneity inherent in distributed systems.

3. Cluster computing is a type of parallel or distributed processing system, which consists of a collection of _____ stand-alone computers cooperatively working together as a single, integrated computing resource.

4. Grid computing aims to "enable _____ and coordinated problem solving in dynamic, multi-institutional virtual organizations".

5. Cloud computing is hinting at a future in which we won't compute on local computers, but on centralized facilities operated by _____ compute and storage utilities.

II. Translate the following terms or phrases from English into Chinese.

information retrieval   distributed system
fault tolerance   cluster computing
load-balancing   grid computing
cloud computing

Section C Software Configuration Guide For Cisco 2600 Series Routers

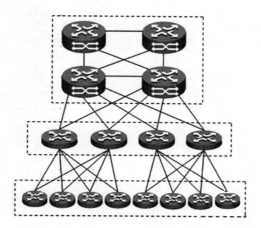

Understanding Interface Numbering and Cisco IOS Software Basics

This chapter provides an overview of the **interface** numbering in the Cisco① 2600 series routers. It also describes how to use the Cisco IOS software are commands.

Understanding Interface Numbering

This section contains information with which you should be familiar before you begin to configure your router for the first time, including interface numbering and what you should do before starting your router.

Cisco 2600 Series Interface Numbering

Each network interface on a Cisco 2600 series router is identified by a **slot** number and a unit number. Table 5C-1 lists the router models and summaries the interfaces supported on each model that are available in the Cisco 2600 series routers.

---

① Cisco：思科公司，全球领先的网络解决方案供应商。

Table 5C-1 Summary of Cisco 2600 Series Router and Interface

Model	Cisco 2610	Cisco 2610XM	...
Ethernet(10BASE-T)	1		
Token-Ring(RJ-45)			
Fast Ethernet(10/100)		1	
Network Module Slot	1	1	
WAN Interface Card Slots	2	2	
Advanced Integration Module Slots	1	1	

Note The number and type of interfaces vary depending on the router.

WAN and LAN Interface Numbering

The Cisco 2600 series router **chassis** contains the following wide-area network(WAN) and local-area network(LAN) interface types:

- **Built-in** LAN interfaces Ethernet, Fast Ethernet, **Token** Ring.
- Two or three slots in which you can install WAN interface cards(WICs).
- One slot in which you can install a network module.

The numbering format is $<interface\text{-}type><Slot\text{-}number><Interface\text{-}number>$. Two examples are:

```
Ethernet 0/0
Serial 1/2
```

The slot number is 0 for all built-in interfaces and 0 for all WIC interfaces; the slot number is 1 for network module interfaces.

Interface(**port**) numbers begin at 0 for each interface type, and continue from right to left and(if necessary) from bottom to top.

Figure 5C-1 below shows a router of 1 RU[①] height with:

[①] RU 是 Rack Unit 的缩略词，U 是由美国电子工业联盟定义的机架的基本高度单位。1U＝1.75 in(4.445 cm)，1RU 指设备的外形满足 EIA 规格，即宽 19 in(48.26 cm)，厚 1.75 in(4.445 cm)。

chassis
/'ʃæsɪ/
n. 底盘；底架
built-in
嵌入的；内置
token
/'təʊkən/
n. 令牌

serial
/'sɪərɪəl/
n. 串口；串行
port
/pɔːt/
n. 端口

- A WIC in each WIC slot (containing interface Serial 0/0 in physical slot W0, and interface Serial 0/1 in physical slot W1).
- A 4-serial-port network module in slot 1 (containing the following ports: Serial 1/0, Serial 1/1, Serial 1/2, and Serial 1/3).
- First built-in Ethernet interface—Ethernet 0/0.
- Second built-in Ethernet interface—Ethernet 0/1, or optionally in Cisco 2612 and Cisco 2613 only: Token Ring interface 0/0.

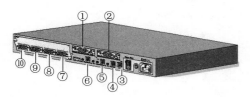

1	Serial 0/1	6	Ethernet 0/1
2	Serial 0/1	7	Serial 1/0
3	Auxiliary port	8	Serial 1/1
4	Console port	9	Serial 1/2
5	Ethernet 0/0	10	Serial 1/0

Figure 5C-1　Example of 1RU Router

Note　The slot number for all WIC interfaces is always 0 (The W0 and W1 slot designations are for physical slot identification only). Interfaces in the WICs are numbered from right to left, starting with 0/0 for each interface type, regardless of which physical slot the WICs are installed in.

Some examples are:
- If physical slot W0 is empty and physical slot w1 contains a 1-port serial WIC, the interface number in the WIC is numbered Serial 0/0.
- If slot W0 contains a 2-port serial WIC and slot W1 contains a 1-port serial WIC, the interfaces in physical slot W0 are numbered Serial 0/0 and Serial 0/1, and the interface in physical slot W1 is numbered Serial 0/2.
- If slot W0 contains a 2-port serial WIC and slot W1 contains a 1-port BRI WIC, the interfaces in physical slot W0 are numbered Serial 0/0 and Serial 0/1, and the interface in physical slot W1 is numbered BRI 0/0.

Voice Interface Numbering in Cisco 2600 Series Routers

Voice interface are numbered differently from the WAN interfaces described in the previous section.

Voice interfaces are numbered as follows:

<Chassis slot><voice module slot<voice interface>

If a 4-**channel** voice network module is installed in chassis slot 1, the voice interfaces are

- 1/0/0—Chassis slot 1/Voice module slot 0/Voice interface 0
- 1/0/1—Chassis slot 1/Voice module slot 0/Voice interface 1
- 1/1/0—Chassis slot 1/Voice module slot 1/Voice interface 0
- 1/1/1—Chassis slot 1/Voice module slot 1/Voice interface 1

channel
/'tʃænl/
n. 通道;频道

Understanding Cisco IOS Software Basics

This section describes what you need to know about the Cisco IOS software before you configure the router using the command-line interface(**CLI**).

CLI
命令行界面

Getting Help

Use the question mark(?) and **arrow keys** to help you enter commands:

- For a list of available commands, enter a question mark:

  Router> ?

- To complete a command, enter a few known characters followed by a question mark(with no **space**):

  Router>s?

- For a list of command **variable**, enter the command followed by a space and a question mark:

  Router>show?

- To redisplay a command you previously entered, press the up arrow key. You can continue to press the up arrow key for more command.

arrow key
箭头键

space
/speɪs/
n. 空白

variable
/'veərɪəbl/
n. 变量

Understanding Command Modes

The Cisco IOS user interface is divided into different modes. Each command mode permits you to configure different components on your router. The commands available at any given time depend on which mode you are currently in. Entering a question mark(?) at the **prompt**, displays a list of commands available for each command mode. Table 5C-2 lists the most common command modes.

prompt
/prɒmpt/
n. 提示符

Table 5C-2 Common Command Modes

Command Mode	Access Method	Router Prompt Displayed	Exit Method
User EXEC	Log in	Router>	Use the **logout** command
Privileged EXEC	From user EXEC mode, enter the enable command	Router#	To exit to user EXEC mode, use the **disable**, exit, or **logout** command
Global configuration	From the privileged EXEC mode, enter the **configure terminal** command	Router(config)#	To exit to privileged EXEC mode, use the **exit** or **end** command, or press **Ctrl-Z**
Interface configuration	From the global configuration mode, enter the **in-terface**<*type number*> command, such as **interface serial 0/0**	Router(config-if)#	To exit to global configuration mode, use the **exit** command. To exit directly to privileged EXEC mode, press **Ctrl-Z**

In the following example, notice how the prompt changes after each command to indicate a new command mode:

```
Router>enable
Password: <enable password>
Router# configure terminal
Router(config)# interface serial 0/0
Router(config-if)# line 0
Router(config-line)# controller t1 0
Router(config-controller)# exit
Router(config)# exit
Router#
% SYS-5-CONFIG_I: Configured from console by console
```

console
/kən'səʊl/
n. 控制台

The last message is normal and does not indicate an error.

Press **Return** to get the Router# prompt.

Note You can press **Ctrl-Z** in any mode to immediately return to enable mode(Router#), instead of entering **exit**, which returns you to the previous mode.

Undoing a Command or Feature

If you want to undo a command you entered or disable a feature, enter the keyword no before most commands; for example, **no ip routing**.

Saving Configuration Changes

You need to enter the **copy running-config startup-config** command to save your configuration changes to **nonvolatile** random-access memory(NVRAM), so the change are not lost if there is a system **reload** or power **outage**. For example:

```
Router# copy running-config startup-config
Building configuration...
```

It might take a minute or two to save the configuration to NVRAM. After the configuration has been saved, the following appear:

```
[OK]
Router#
```

nonvolatile
/nɒnˈvɒlətaɪl/
adj. 非易失性的

reload
/ˌriːˈləʊd/
n. 重新加载

outage
/ˈaʊtɪdʒ/
n. 断电;停机

Exercises

Ⅰ. **Fill in the blanks with the information given in the text.**

1. Each network interface on a Cisco 2600 series router is identified by a _____ number and a unit number.

2. The numbering format is _____ <Slot-number><Interface-number>.

3. Interface numbers begin at _____ for each interface type, and continue from _____ to left and (if necessary) from bottom to top.

4. To redisplay a command you previously entered, press the _____ arrow key.

5. If you want to undo a command you entered or disable a feature, enter the keyword _____ before most commands.

II. Translate the following terms or phrases from English into Chinese.

IOS	built-in
fast ethernet	token ring
WAN interface card	command-line interface(CLI)

Unit Six Internet

Section A Internet

The Internet is a global system of interconnected computer networks that use the standard Internet protocol **suite**(TCP/IP) to link several billion devices worldwide. It is an international network of networks that consists of millions of private, public, **academic**, business, and government **packet switched** networks, linked by a broad array of electronic, wireless, and optical networking technologies. The Internet carries an extensive range of information resources and services, such as the inter-linked **hypertext** documents and applications of the **World Wide Web**(WWW), the infrastructure to support E-mail, and peer-to-peer networks for file sharing and telephony.

Ⅰ. History

When the Soviets launched **Sputnik** satellite in space in 1957, American started agency names Advance Research Project Agency (ARPA①). In order to maintain communication with its computer network even if part of the network was destroyed during an attack, ARPA started funding the design of the Advanced Research

---

① ARPA: Advance Research Project Agency,美国高级研究计划局,后更名为美国国防高级研究计划局(DARPA),负责研发用于军事用途的高新科技。互联网、半导体、GPS、激光器等许多重大科技成果都出自于它。

Projects Agency Network (ARPAnet①) for the United States Department of Defense②. ARPAnet, which only four computers connected, was officially launched in 1969. This was the predecessor to the Internet.

By the 1970s, ARPAnet had dozens of computer networks, but each network could only communicate with other computers inside the network. Different computer networks were still unable to communicate with each other. In response, ARPA had set up a new research project to connect different computer networks in a new way to form the "Internet". Researchers call this internetwork. In 1974, protocols were developed to connect **packet networks**, including TCP/IP proposed by Vinton. Cerf and Bob Braden. ARPA accepted TCP/IP in 1982 and chose the Internet as the primary computer communication system.

packet network
分组网络

Ⅱ. Development

In 1986, the National Science Foundation(NSF③) of the United States connected five supercomputer centers for research and education across the United States to form NSFnet. In 1988, NSFnet replaced ARPAnet as the **backbone** of the Internet.

In 1989, Tim Berners Lee④ designed the first web server and the first client. He invented the first web browser for the NeXT computer⑤. The browser's name was World Wide Web.

The development of Internet had aroused great interest in business. In 1992, IBM, MCI⑥ and MERIT⑦ jointly established an advanced network service Company(ANS) and established a new network called ANSnet, which became another backbone network of the Internet, thus making the Internet began to commercialize.

backbone
/ˈbækbəʊn/
n. 骨干

---

① ARPAnet: the Advanced Research Projects Agency Network,高级研究计划局网。
② the United States Department of Defense:美国国防部。
③ NSF: the National Science Foundation,国家科学基金组织。
④ Tim Berners Lee: 蒂姆·伯纳斯·李,英国计算机科学家,万维网的发明者,他发明了第一个网页浏览器,创办了万维网联盟(W3C)。
⑤ NeXT computer: 1988 年由 NeXT 公司(史蒂夫·保罗·乔布斯创办)推出的第一个工作站计算机产品。
⑥ MCI: Microwave Communications Inc.,美国电信公司。
⑦ MERIT:美国网络公司,全球领先、历史悠久的国际性知名信息维护外包公司。

In April 1993, **Mosaic** was released by the NCSA① at the University of Illinois② which was developed by Marc Andreessen③, a student, and his friends. It was the first browser to be used on the Internet. Marc Andreessen founded **Netscape**, they completely rewrote Mosaic's code, and the browser's name was changed to Navigator. It accelerated the rapid growth of the Web.

On April 30, 1995, NSFnet officially announced its closure. At this time, the backbone of the Internet had covered 91 countries around the world, with more than 4 million hosts. In recent years, the Internet has developed at an amazing speed.

Ⅲ. Protocol

While the Internet's hardware can often be used to support other software systems, it is the design and the standardization process of the software architecture that characterizes the Internet and provides the foundation for its scalability and success. The responsibility for the architectural design of the Internet software systems has been assumed by the Internet Engineering Task Force (**IETF**④). The IETF conducts standard-setting work group, open to any individual, about the various aspects of Internet architecture. Resulting discussions and standards are published in a series of publications, each called a Request for Comments(**RFC**⑤), on the IETF web site.

The principal methods of networking that enable the Internet are contained in specially designated RFCs that constitute the Internet Standards. Other less **rigorous** documents are simply informative, experimental, or historical, or document the best current practices when implementing Internet technologies.

The Internet standards describe a framework known as the Internet protocol suite. This is a model architecture that divides methods into a layered system of protocols, originally documented

---

① NCSA：National Center for Supercomputing Applications,国家超级计算应用中心。
② the University of Illinois：伊利诺伊大学,位于美国伊利诺伊州。
③ Marc Andreessen：马克·安德森(1971—   ),美国人,开发了第一个浏览器 Mosaic。
④ IETF：互联网工程任务组,是互联网最具权威的技术标准化组织。
⑤ RFC：请求评议,绝大部分网络标准的指定都是以 RFC 的形式开始,经过大量的论证和修改,最终由主要的标准化组织指定。

in RFC 1122 and RFC 1123. The layers correspond to the environment or scope in which their services operate. Figure 6A-1 shows the TCP/IP archtecture.

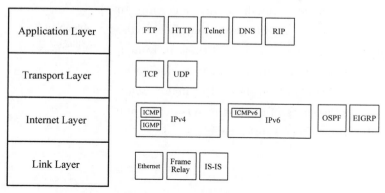

Figure 6A-1　TCP/IP Archtecture

At the top is the **application layer**, the space for the application-specific networking methods used in software applications. For example, a web browser program uses the client-server application model and a specific protocol of interaction between servers and clients, while many file-sharing systems use a peer-to-peer paradigm. Below this top layer, the **transport layer** connects applications on different **hosts** with a logical channel through the network with appropriate data exchange methods.

Underlying these layers are the networking technologies that interconnect networks at their borders and hosts via the physical connections. The **internet layer** enables computers to identify and locate each other via internet protocol(**IP**) addresses, and routes their traffic via intermediate (transit) networks. Last, at the bottom of the architecture is the **link layer**, which provides connectivity between hosts on the same network link, such as a physical connection in form of a local area network(LAN) or a dial-up connection. The model, also known as TCP/IP, is designed to be independent of the underlying hardware, which the model therefore does not concern itself with in any detail. The link layer is also called the **network access layer**.

Ⅳ. Application

Most traditional communications media including telephone,

music, film, and television are being reshaped or redefined by the Internet, giving birth to new services such as voice over Internet Protocol(**VoIP**①) and Internet Protocol television(**IPTV**②). Newspaper, book, and other print publishing are adapting to website technology, or are reshaped into **blogging** and web feeds. The Internet has enabled and accelerated new forms of human interactions through instant messaging, Internet forums, and social networking. Online shopping has boomed both for major retail outlets and small artisans and traders. Business-to-business③ and financial services on the Internet affect supply chains across entire industries.

blogging
/ˈblɒɡɪŋ/
n. 博客

Ⅴ. Internet Plus

The Internet Plus④ action plan was **unveiled** in the government work report that Premier Li Keqiang of **the State Council** delivered to National People's Congress(NPC⑤) on March 5, 2015. Li said, China will develop the "Internet Plus" action plan to integrate mobile Internet, cloud computing, big data and the Internet of things with modern manufacturing, to encourage the healthy development of E-commerce, industrial networks, and Internet banking, and to help Internet companies increase their international presence.

Internet Plus is the integration of the Internet and traditional industries through online platforms and IT technology, it is expected to help economic restructuring, improve people's **livelihoods** and transform of government functions. The Internet Plus action plan change our life in all different parts.

Internet Plus can be both challenge and opportunity. In the recent several years, some education-related Apps have emerged

unveiled
/ˌʌnˈveɪld/
adj. 公布于众的
the State Council
国务院

livelihood
/ˈlaɪvlihʊd/
n. 民生；生计

---

① VoIP：称为 IP 电话、网络电话，将模拟信号数字化，以数据封包的形式在 IP 网络上做实时传递。
② IPTV：即交互式网络电视，是一种利用宽带网，集互联网、多媒体、通信等技术于一体，向家庭用户提供包括数字电视在内的多种交互式服务的崭新技术。
③ business-to-business：电子商务的一种类型，即企业—企业电子商务。
④ Internet Plus：互联网＋，2015 年 3 月 5 日，第十二届全国人民代表大会第三次会议上，李克强总理政府工作报告中提出制定"互联网＋"行动计划。
⑤ NPC：全国人民代表大会。

and become increasing popular. For example, MOOC① provides students a profusion of courses all around the world for free. Internet plus transportation, agriculture, city management, industrial monitoring, medical, environmental monitoring, etc., and they provide us with challenge and opportunity and we need make good use of it.

Exercises

Ⅰ. **Fill in the blanks with the information given in the text.**

1. The Internet is a global system of _____ computer networks.
2. The Internet carries an extensive range of information resources and services, such as the inter-linked _____ documents and applications of the _____, the infrastructure to support E-mail, and peer-to-peer networks for file sharing and telephony.
3. The origins of the Internet date back to research commissioned by the United States government in the _____.
4. The principal methods of networking that enable the Internet are contained in specially designated _____ that constitute the Internet Standards.
5. The internet layer enables computers to identify and locate each other via _____ addresses, and routes their traffic via intermediate (transit) networks.
6. _____ provides students a profusion of courses all around the world for free.

Ⅱ. **Translate the following terms or phrases from English into Chinese.**

TCP/IP	protocol suite
World Wide Web(WWW)	hypertext
packet switched network	peer-to-peer network
IETF	RFC
host	Internet Protocol(IP)
VoIP	web site
IPTV	business-to-business
MOOC	

---

① MOOC: Massive Open Online Courses, 慕课, 大规模开放在线课程。

III. Translate the following passage from English into Chinese.

Smarthome

Smarthome is one of the world's largest home automation retailers, becoming an easy-to-use source for affordable devices—including smart lighting control, smart thermostats, smart home security, wireless cameras, doorbell cameras, door locks, and much more—all of which the average do-it-yourselfer can safely install.

Insteon is the gold standard in wireless networking technology for the connected home. It features lighting control, scene lighting, and timers; along with leak, door and motion sensing alerts—all from your smartphone or tablet. Insteon-compatible products are affordable, easy-to-program devices that communicate over your home's existing wiring and over wireless signals, making it a dual-mesh system, that significantly increases signal reliability.

Section B　5G

Ⅰ. Introduction

5G(fifth-generation mobile communications) is a new generation of mobile communication systems for 2020, with high **spectral** efficiency and low power consumption, in terms of transfer rate and resource utilization improvement over 4G system 10 times, its wireless coverage performance and user experience will be significantly improved. 5G will be closely integrated with other wireless mobile communication technology, constitute a new generation of ubiquitous mobile information network, to meet future mobile Internet **traffic** 1000x development needs in 10 years.

Ⅱ. Research and Development History

February 2013, **the EU** announced that it would grant 50 million **euros** to accelerate the development of 5G mobile technology, plans to launch a mature standard in 2020.

May 13, 2013, South Korea's Samsung Electronics Co., Ltd. announced that it has successfully developed the 5th generation mobile communication(5G) core technology, which is expected to begin in 2020 to **commercialization**. The technology can transmit data at ultra-high frequency 28GHz to 1Gb/s, and the maximum transmission distance of up to 2 km.

Back in 2009, Huawei has launched the early research related technologies, and to show the prototype of the 5G base in later

spectral
/ˈspektrəl/
adj. 光谱的

traffic
/ˈtræfɪk/
n. 流量

EU
欧盟
euro
/ˈjʊərəʊ/
n. 欧元

commercialization
/kəˌmɜːrʃələˈzeɪʃn/
n. 商业化;商品化

years. In November 6, 2013, Huawei announced that it would invest $600 million in 2018 for the 5G technology development and innovation, and predicted that users will enjoy 20Gb/s commercial 5G mobile networks in 2020.

May 8, 2014, the Japanese telecom operator NTT DoCoMo① announced officially, Ericsson②, Nokia③, Samsung④ and other six manufacturers to work together, began testing override 1000 times than existing 4G networks the carrying capacity of the high-speed network 5G network, the transmission speed is expected to 10Gb/s. Outdoor testing scheduled to commence in 2015, and expects to begin operations in 2020.

A number of countries and organizations announced, 5G network will be operational between 2020-2025.

Ⅲ. Core Concept

What is 5G? 5G refers to the fifth generation of mobile communications. Currently, the global industry for 5G concept not yet agreed. China IMT-2020⑤ (5G) group released the White Paper⑥ considers the concept 5G, 5G integrated key capabilities and core technology, 5G concept by "important targets" and "a group of key technologies" to a common definition. Among them, the flag indicators "Gb/s rate user experience" is a set of key technologies, including large-scale **antenna** array, **ultra-dense** networking, new multi-site, full-spectrum access and new network architectures.

Unlike the case in the past only to emphasize different **peak rate**, the industry generally believe that the rate of the user experience is the most important performance indicators, it truly reflects the real data rate available to the user, and the user

antenna
/æn'tenə/
n. 天线
ultra-dense
超高密度
peak rate
峰值速率

---

① NTT DoCoMo：日本的一家电信公司，是日本最大的移动通信运营商。
② Ericsson：爱立信公司。成立于瑞典，公司业务有通信网络系统、电信服务等。
③ Nokia：诺基亚公司。总部位于芬兰埃斯波，主营移动通信设备生产和相关服务的跨国公司。
④ Samsung：三星集团。韩国最大的跨国企业集团，业务涉及电子、金融等领域。
⑤ IMT-2020：2015年10月，在瑞士日内瓦召开的2015无线电通信全会上，国际电联无线电通信部门正式批准了三项有利于推进未来5G研究进程的决议，并正式确定了5G的法定名称是"IMT-2020"。
⑥ the White Paper：白皮书。政府或议会正式发表的以白色封面装帧的重要文件或报告书的别称。

experience is the closest performance.

IV. Challenges

- **Transport Challenge**

We can define 5G in terms of scenarios which the next generation wireless access networks will have to support. A total of five future scenarios have been defined, namely amazingly fast, great service in a crowd, ubiquitous things communicating, super real time and reliable connections, and best experience follows you.

Three of these challenges (i.e., very high data rate, very dense crowds of users and mobility) are more traditional in the sense that they are related to continued enhancement of user experience and supporting increasing traffic volumes and mobility. Two emerging challenges, very low **latency** and very low energy, cost and massive number of devices, are associated with the application of wireless communications to new areas. Future applications may be associated with one or several of these scenarios imposing different challenges to the network.

Support for very high data rates will require both higher capacity radio access nodes as well as a densification of radio access sites. This, in turn, translates into a transport network that needs to support more sites and higher capacity per site, i.e. huge traffic volumes. The great service in a crowd scenario will put requirements on the transport network to provide very high capacity on-demand to specific geographical locations. In addition, the best experience follows your scenario, suggests a challenge in terms of fast reconfigurability of the transport resources. On the contrary, the other 5G challenges are not expected to play as important role for shaping the transport, as for example the case of very low latency and very low energy, cost and massive number of devices. A properly dimensioned transport network based on modern wireless and/or optical technologies is already today able to provide extremely low latency, i.e., the end-to-end delay contribution of the transport network is usually almost **negligible**. In addition, while a huge number of connected machines and devices will create a challenge for the wireless network, it will most

latency
/ˈleɪtənsi/
n. 延迟；时延

negligible
/ˈneɡlɪdʒəbl/
adj. 微不足道的

probably not significantly impact the transport. This is due to the fact that the traffic generated by a large number of devices over a geographical area will already be aggregate graphical area will already be **aggregated** in the transport.

- **Machine to Machine Communication**

Besides network evolution, we observe also device evolution that become more and more powerful. The future wireless landscape will serve not only mobile users through such devices as smartphones, tablets or game consoles but also a tremendous number of any other devices, such as cars, smart grid terminals, health monitoring devices and household appliances that would soon require a connection to the Internet. The number of connected devices will **proliferate** at a very high speed. It is estimated that the M2M traffic increased 24-fold between 2012 and 2017.

- **Core Network Virtualisation**

Moving towards 5G imposes changes not only in the Radio Access Network(**RAN**) but also in the Core Network(**CN**), where new approaches to network design are needed to provide connectivity to growing number of users and devices. The trend is to **decouple** hardware from software and move the network functions towards the latter one. Software Defined Networking (SDN①) being standardised by Open Networking Foundation (**ONF**) assumes separation of the control and data plane. Consequently, thanks to centralization and programmability, configuration of forwarding can be greatly automated.

Moreover, standardization efforts aiming at defining Network Functions Virtualisation(**NFV**) are conducted by multiple industrial partners including network operators and equipment vendors within ETSI②. Introducing a new software based solution is much faster than installing an additional specialised device with a particular functionality. Both solutions would improve the network adaptability and make it easily scalable.

aggregate
/ˈæɡrɪɡət/
v. 聚合；聚集

proliferate
/prəˈlɪfəreɪt/
v. 激增

RAN
无线接入网络
CN
核心网络
decouple
/diːˈkʌpl/
v. 使分离
ONF
开放网络基金会

NFV
网络功能虚拟化

---

① SDN：软件定义网络，美国斯坦福大学提出的一种新型网络架构，以实现网络虚拟化。其核心技术 OpenFlow 将网络设备的控制面与数据面分离开，以实现网络流量的灵活控制，使网络作为管道变得更加智能。

② ETSI：European Telecommunications Standards Institute，欧洲电信标准化协会。

There are a lot of unresolved issues before 5G networks are truly operational. Also faced include how to design network architecture, including many technical challenges. Compared with previous generations of communications technology, 5G era's biggest challenge is not how to increase the rate, but the user experience with traffic density, the number of terminals from a series of **interwoven** problems. As much as possible while also reducing user costs. This is the 5G network must be solved.

interwoven
/ˌɪntəˈwəʊvən/
v. 交织

Ⅴ. 5G in China

February 11, 2015, China released White Paper about concept of 5G. It instantly make more people are concerned about 5G.

China will actively participate in the development of 5G standards, will help China to further enhance the **patent** position in international communication standards, **escort** for our mobile phone manufacturing.

patent
/ˈpæt(ə)nt/
n. 专利

escort
/ˈeskɔːt/
v. 保驾护航

It is worth noting that, in the 5G standards, Huawei, ZTE① and other Chinese telecommunications companies such as Ericsson (China) communications Co.LTD② also play an important role, in which Huawei from between 2013 to 2018, five years is ho throw $ 600 million 5G conduct research and innovation.

China needs to have its own place in the 5G market, China's communications companies are also very hard, believe in the future, China's R&D③ level 5G will lead other countries.

Exercises

Ⅰ. Fill in the blanks with the information given in the text.

1. 5G is a new generation of _____ communication systems.

2. Back in 2009, Huawei has launched the early research related technologies, and to show the _____ of the 5G base in later years.

3. Huawei predicted that users will enjoy _____ commercial 5G mobile networks

---

① ZTE：中兴通讯股份有限公司,全球领先的综合通信解决方案提供商,中国最大的通信设备上市公司。
② Ericsson(China) communications Co.LTD：爱立信(中国)通信有限公司。
③ R&D：Research & Developing,研发。

in 2020.

4. China IMT-2020(5G) group released the White Paper, among them, the flag indicators "_____" is a set of key technologies, including large-scale antenna array, ultra-dense networking, new multi-site, full-spectrum access and new network architectures.

5. Two emerging challenges, very low _____ and very low _____, cost and massive number of devices, are associated with the application of wireless communications to new areas.

6. The trend is to decouple hardware from _____ and move the network functions towards the latter one.

7. China will actively participate in the development of 5G _____, will help China to further enhance the patent position in international communication standards.

II. Translate the following terms or phrases from English into Chinese.

5G
low power consumption
wireless coverage performance
low latency
the White Paper
peak rate

high spectral efficiency
transfer rate
user experience
SDN
ultra-dense

Section C　Top 10 Search Engines in the World

Which are the 10 best and most popular search engines in the World? Besides Google and Bing, there are other search engines that may not be so well known but still serve millions of search queries per day.

The 10 best search engines in 2022 ranked by popularity are: Google, Microsoft Bing, Yahoo, Baidu, Yandex, DuckDuckGo, Ask, Ecosia, Aol, Internet Archive.

Ⅰ. Google

Google is the most popular search engine with a **stunning** 91.42% market share. It holds first place in search with a difference of 88.28% from second in place Bing. According to statistics from Statista and StatCounter, Google is dominating the market in all countries on any device(desktop, mobile, and tablet).

What made Google the most popular and trusted search engine is the quality of its search results. Google is using sophisticated algorithms to present the most accurate results to the users.

Google's founders Larry Page① and Sergey Brin② came up with the idea that websites referenced by other websites are more important than others and thus deserve a higher ranking in the

stunning
/ˈstʌnɪŋ/
adj. 令人震惊的

---

① Larry Page: 拉里·佩奇(1973—), 谷歌搜索引擎的创始人之一。
② Sergey Brin: 谢尔盖·布林(1973—), 谷歌公司联合创始人之一。

search results.

Over the years the Google ranking algorithm has been enriched with hundreds of other factors (including the help of machine learning) and still remains the most reliable way to find exactly what you are looking for on the Internet.

II. Microsoft Bing

Bing[①] was renamed Microsoft Bing in October 2020.

The best alternative search engine to Google is Microsoft Bing. Bing's search engine share is between 2.83% and 12.31%.

Bing is Microsoft's attempt to challenge Google in search, but despite their efforts, they still did not manage to convince users that their search engine can be as reliable as Google. Their search engine market share is constantly low even though Bing is the default search engine on Windows PCs.

Bing originated from Microsoft's previous search engines (MSN Search, Windows Live Search, Live Search), and according to Alexa[②] rank is the 30th most visited website on the Internet.

III. Yahoo

Yahoo is one of the most popular email providers and its web search engine holds third place in search with an average of 1% market share.

From October 2011 to October 2015, Yahoo search was powered exclusively by Bing. In October 2015 Yahoo agreed with Google to provide search-related services and until October 2018, the results of Yahoo were powered both by Google and Bing. As of October 2019, Yahoo Search is once again provided exclusively by Bing.

Yahoo is also the default search engine for Firefox[③] browsers in the United States since 2014.

---

① Bing：微软公司于 2009 年 5 月推出的搜索引擎，中文品牌名为"必应"，2020 年改名为 Microsoft Bing。
② Alexa：一家专门发布网站世界排名的网站，是当前拥有 URL 数量最庞大、排名信息发布最详尽的网站。
③ Firefox：中文名为"火狐"，由 Mozilla 开发的自由及开放源代码的网页浏览器。

Ⅳ. Baidu

Baidu① has a global market share between 0.68% and 11.26%.

Baidu was founded in 2000 and it is the most popular search engine in China. Its market share is increasing steadily and according to Wikipedia②. Baidu is serving billions of search queries per month. It is currently ranked at position 4, in the Alexa Rankings.

Although Baidu is accessible worldwide, it is only available in the Chinese language.

Ⅴ. Yandex

Yandex③, Russian's most popular search engine has a global market share between 0.5% and 1.16%.

According to Alexa, Yandex is among the 30 most popular websites on the Internet with a ranking position of 4 in Russian.

Yandex presents itself as a technology company that builds intelligent products and services powered by machine learning.

According to Wikipedia, Yandex operates the largest search engine in Russia with about 65% market share in that country.

Ⅵ. DuckDuckGo

DuckDuckGo's④ search engine market share is around 0.66%.

According to DuckDuckGo traffic statistics, they are serving on average 90 million searches per day but still their overall market share is constantly below 0.6%.

Unlike what most people believe, DuckDuckGo does not have a search index of their own(like Google and Bing) but they generate

---

① Baidu,中文名为"百度",公司创始人李彦宏于2000年1月1日于中关村创立了百度公司,百度是拥有强大互联网基础的领先AI公司。
② Wikipedia:中文名为"维基百科",总部位于美国,是一个基于维基技术,用多种语言编写而成的网络百科全书。
③ Yandex:俄罗斯重要的网络服务门户之一,俄罗斯网络拥有用户最多的网站。
④ DuckDuckGo:互联网搜索引擎,总部位于美国宾州Valley Forge市。其办站哲学主张维护使用者的隐私权,并承诺不监控、不记录使用者的搜寻内容。

their search results using a variety of sources. In other words, they don't have their own data but they depend on other sources (like Yelp①, Bing, Yahoo) to provide answers to users' questions. This is a big limitation compared to Google that has a set of algorithms to determine the best results from all the websites available on the Internet.

On the positive side, DuckDuckGo has a clean interface, it does not track users and it is not fully loaded with advertisings.

Ⅶ. Ask

Formerly known as AskJeeves, Ask.com receives approximately 0.42% of the search share. ASK is based on a question/answer format where most questions are answered by other users or are in the form of polls.

It also has the general search functionality but the results returned lack quality compared to Google or even Bing and Yahoo.

Ⅷ. Ecosia

Ecosia② is a Berlin-based social business founded by Christian Kroll in 2009. The main reason Ecosia was created was to help in financing planting trees and restoration projects. It is thus known as the "tree planting search engine".

How does Ecosia work? Ecosia is a Bing partner, meaning that its search results are powered by Bing. Ecosia makes money to support the planning of trees by displaying ads in their search results. Every time an ad is clicked, Ecosia gets a small share. It is estimated that it takes approx 45 searches to finance the planting of one tree. In terms of search engine market share, Ecosia's share is around 0.10%.

---

① Yelp：美国最大的点评网站。
② Ecosia：一款基于微软 Bing 和雅虎搜索引擎功能的全新搜索引擎。在用户访问过程中，单击产生的广告费收入，80%捐赠于世界自然基金会（WWF），以开展热带雨林保护项目，应对全球变暖，被称为"绿色搜索引擎"。

Ⅸ. AOL

The old-time famous AOL① is still in the top 10 search engines with a market share that is close to 0.05%. The AOL network includes many popular websites like engadget.com, techchrunch.com, and huffingtonpost.com. On June 23, 2015, AOL was acquired by Verizon Communications.

Ⅹ. Internet Archive

archive
/'ɑːkaɪv/
n. 存档

Archive.org is the internet archive search engine. You can use it to find out how a web site looked since 1996. It is a very useful tool if you want to trace the history of a domain and examine how it has changed over the years.

Exercises

Ⅰ. Fill in the blanks with the information given in the text.

1. Google is the most popular and trusted search engine because of the _____ of its search results.
2. The best alternative search engine to Google is _____.
3. Yahoo is the default search engine for _____ browsers in the United States.
4. Although Baidu is accessible worldwide, it is only available in the _____ language.
5. _____ is known as the "tree planting search engine".

Ⅱ. Translate the following terms or phrases from English into Chinese.

search engine	Microsoft Bing
MSN	Firefox browser
Wikipedia	traffic statistics

---

① AOL：即 American Online，美国在线，是美国最大的因特网服务提供商之一。

Unit Seven The Internet of Things

Section A Cloud Infrastructure and Services

I. Introduction

For organizations to be competitive in today's fast-paced, online, and highly interconnected global economy, they must be **agile**, flexible, and able to respond rapidly to the changing market conditions. Cloud, a next generation style of computing, provides highly scalable and flexible computing that is available on demand.

Historically, **cloud computing** has evolved through grid computing, utility computing, **virtualization**, and service oriented architecture(**SOA**). NIST[①] has defined cloud computing as follows: A model for enabling **ubiquitous**, convenient, on-demand network access to a shared pool of configurable computing resources (e.g. servers, storage, networks, applications, and services) that can be rapidly **provisioned** and released with minimal management effort or service provider interaction. Also the cloud infrastructure should essentially have five essential characteristics:

---

① NIST: 美国国家标准与技术研究院(National Institute of Standards and Technology)。

- **On-Demand Self-Service**

The on-demand and self-service aspects of cloud computing mean that a consumer can use cloud services as required, without any human intervention with the cloud service provider.

- **Broad Network Access**

Cloud services are accessed via the network, usually the Internet, from a broad range of client platforms, such as desktop computer, laptop, mobile phone, and thin client. Traditionally, softwares such as Microsoft Word or Microsoft PowerPoint, have been offered as client-based software. Users have to install the software on their computers in order to use this software application.

- **Resource Pooling**

A cloud must have a large and flexible resource pool to meet the consumer's needs, to provide the economies of scale, and to meet service-level requirements. The resources (compute, storage, and network) from the pool are dynamically assigned to multiple consumers based on a multi-**tenant** model.

- **Rapid Elasticity**

Refers to the ability of the cloud to expand or reduce **allocated IT** (Information Technology) resources quickly and efficiently. This allocation might be done automatically without any service interruption. The cloud enables to grow and **shrink** these resources dynamically and allows the organizations to pay on a usage basis.

- **Metered Service**

The metered services continuously monitor resource usage (CPU time, bandwidth, storage capacity) and report the same to the consumer. It provides billing and chargeback information for the cloud resource used by the consumer.

II. Cloud Infrastructure

An infrastructure should fulfill the essential characteristics to

tenant
/'tenənt/
n. 租户
elasticity
/ˌelæs'tɪsəti/
n. 弹性布署
allocate
/'æləkeɪt/
vt. 分配
IT
信息技术
shrink
/ʃrɪŋk/
v. 减少
metered
/'mi:təd/
adj. 计量的

support cloud services. See Figure 7A-1 as bellows.

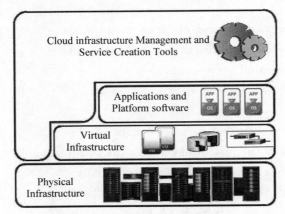

Figure 7A-1　Cloud Infrastructure

- **Physical Infrastructure**

The physical infrastructure consists of physical IT resources that include physical servers, storage systems, and physical network components, such as physical adapters, switches, and routers. Physical servers are connected to each other, to the storage systems, and to the clients via physical networks such as IP network, FC SAN, IP SAN, or FCoE① network.

Cloud service providers may use physical IT resources from one or more data centers to provide services. If the physical IT resources are distributed across multiple data centers, connectivity must be established among them. The connectivity enables data centers in different locations to work as single large data center. This enables both **migration** of cloud services across data centers and provisioning cloud services using resources from multiple data centers.

- **Virtual Infrastructure**

Virtual infrastructure consists of the following resources:
Resource pools such as CPU pools, memory pools, network bandwidth pools, and storage pools.
Identity pools such as **VLAN** ID pools, **VSAN** ID pools, and

migration
/maɪˈgreɪʃn/
n. 迁移
virtual
/ˈvɜːtʃʊəl/
adj. 虚拟的
VLAN
虚拟局域网
VSAN
虚拟存储区域网络

---

① FCoE：Fibre Channel over Ethernet，基于以太网的光纤通信技术网络。

MAC[①] address pools.

Virtual IT resources such as **VMs**, virtual **volumes**, virtual networks, and VM network components such as virtual switches and virtual **NICs**.

- **Applications and Platform Software**

Applications and platform software layers include a suite of softwares such as:
- Business applications.
- Operating systems and database. These softwares are required to build environments for running applications.
- Migration tools.

Applications and platform software are hosted on VMs to create **Software-as-a-Service(SaaS)** and **Platform-as-a-Service(PaaS)**.

For SaaS, applications and platform softwares are provided by the cloud service providers. For PaaS, only the platform software is provided by the cloud service providers; consumers export their applications to cloud. In **Infrastructure as a Service(IaaS)**, consumers upload both applications and platform software to cloud. Cloud service providers supply migration tools to consumers, enabling **deployment** of their applications and platform software to cloud.

- **Cloud Infrastructure Management and Service Creation Tools**

Cloud infrastructure management and service creation tools are responsible for managing physical and virtual infrastructures. They enable consumers to request for Cloud services; they provide Cloud services based on consumer requests and allow consumers to use the services. Cloud infrastructure management and service creation tools automate consumer requests processing and creation of cloud services. They also provide administrators a single management **interface** to manage resources distributed in multiple virtualized data centers(VDCs).

Cloud management tools are classified as:
- Virtual infrastructure management softwares: Enables management of physical and virtual infrastructure resources.

---

① MAC：Mandatory Access Control，介质访问控制层。

> Unified management software: Responsible for creating cloud services.
> User access management software: Enables consumers to request for cloud services.

These softwares interact with each other to automate provisioning of Cloud services.

III. Cloud Service Models

Cloud service models can be classified into three categories: Infrastructure-as-a-Service(IaaS), Platform-as-a-Service(PaaS) and Software-as-a-Service(SaaS).

- **Infrastructure-as-a-Service**

Infrastructure-as-a-Service(IaaS) is the base layer of the cloud stack. It serves as the foundation for the other two layers(SaaS, PaaS) for their execution. The cloud infrastructure such as servers, routers, storage, and other networking components are provided by the IaaS provider. The consumer hires these resources as a service based on needs and pays only for the usage. The consumer is able to deploy and run any software, which may include Operating Systems(OSs) and applications. In fact, IaaS is a mini **do-it-yourself** data center that you would need to configure the resources(server, storage) and to get the job done.

Amazon Elastic Compute Cloud(Amazon EC2)[①] is an Infrastructure-as-a-Service model that provides on-demand scalable compute capacity in the cloud. It enables consumers to leverage Amazon's massive infrastructure with no **up-front** capital investment.

do-it-yourself
自助式

up-front
前期

- **Platform-as-a-Service**

Platform-as-a-Service(PaaS) is the capability provided to the consumer to deploy consumer-created or acquired applications on the cloud infrastructure. PaaS can broadly be defined as application

---

① Amazon EC2: 2006 年 3 月,亚马逊推出了亚马逊弹性可扩展的云计算。它是提供可调节的计算容量的 Web 服务,实际就是 Amazon 数据中心里的服务器,可以使用它来构建和托管用户软件系统。

development environments offered as a "service" by the cloud provider. The consumer uses these platforms that typically have Integrated Development Environment(**IDE**), which includes **editor**, **compiler**, **builder**, and deploy capabilities to develop their applications. For PaaS, consumers pay only for the platform software components such as databases, OS instances, and **middleware**, which includes its associated infrastructure cost.

Google App Engine① is a Platform-as-a-Service that allows consumers to build Web applications using a set of **APIs** and to run those applications on Google's infrastructure. With App Engine, there are no servers to maintain. You just need to **upload** the application, and it is ready to serve.

- **Software-as-a-Service**

Software-as-a-Service(SaaS) is the top most layer of the cloud computing stack, which is directly consumed by the end user. It is the capability, provided to the consumer to use the service provider's applications running on a cloud infrastructure. It is accessible from various client devices through a **thin client** interface such as a Web browser. In a SaaS model, the applications such as **Customer Relationship Management(CRM)**, E-mail, and **Instant Messaging(IM)** are offered as a service by the cloud provider. Here, the consumers will use only the applications they really want and pay a **subscription** fee for the usage. The cloud provider will host and manage the required infrastructure and applications to support these services.

EMC Mozy② is a Software-as-a-Service solution, built on a highly scalable and available **back-end** storage architecture. Consumers can leverage the Mozy **console** to perform automatic, secured, online backup and recovery of their data with ease.

---

① Google App Engine：2008 年 4 月 7 日，谷歌发布了 Google App Engine，让用户可以在谷歌的基础架构上运行自己的网络应用程序，并可根据用户的访问量和数据存储需要的增长轻松扩展。

② EMC Mozy：EMC（易安信）公司创建于 1979 年，是一家全球领先的美国信息存储资讯科技公司，为构建和管理智能、灵活而且安全的信息基础结构提供系统、软件、服务和解决方案。EMC Mozy 是 EMC 推出的云备份。

Ⅳ. Cloud Deployment Model

Cloud computing can be classified into three deployment models: private, public, and **hybrid**. These models provide a basis for how cloud infrastructures are constructed and consumed.

- **Public Cloud**

In a public cloud, IT resources are made available to the general public or organizations and are owned by the cloud service provider. The cloud services are accessible to everyone via standard Internet connections. In a public cloud, a service provider makes IT resources, such as applications, storage capacity, or server compute cycles, available to any consumer. This model can be thought of as an on-demand and as a **pay-as-you-go** environment. However, for organizations, these benefits come with certain risks: no control over the resources in the cloud, the security of **confidential** data, network performance issues, and **interoperability**. Popular examples of public clouds include Amazon's Elastic Compute Cloud(EC2), Google Apps, and Salesforce.com①.

- **Private Cloud**

In a private cloud, the cloud infrastructure is operated solely for one organization and is not shared with other organizations. This cloud model offers the greatest level of security and control.

Like a public cloud, a private cloud also enables provisioning an automated service request rather than a manual task processed by IT. Many enterprises, including EMC, Cisco, IBM, Microsoft, Oracle, and VMware②, now offer cloud platforms and services to build and manage a private cloud.

- **Hybrid Cloud**

In hybrid cloud environment, the organization consumes

---

① Salesforce.com：1999 年由甲骨文（Oracle）高级副总裁、俄罗斯裔美国人马克·贝尼奥夫创办。提出云计算和软件即服务（SaaS）的理念，提供客户关系管理各个方面的按需定制软件服务。

② VMware：1998 年成立，是一个"虚拟 PC"软件公司，在虚拟化和云计算基础架构领域处于全球领先地位，提供服务器、桌面虚拟化的解决方案。

resources from both private and public clouds. The ability to **augment** a private cloud with the resources of a public cloud can be utilized to maintain service levels in the face of rapid **workload fluctuations**. Organizations use their computing resources on a private cloud for normal usage, but access the public cloud for high/peak load requirements. This ensures that a sudden increase in computing requirement is handled gracefully. Ideally, the hybrid approach allows a business to take advantage of the scalability and cost-effectiveness that a public cloud computing environment offers without exposing **mission-critical** applications and data to third-party **vulnerabilities**.

Ⅴ. Cloud Challenges

Both the cloud consumers and providers have their own challenges. The following are the challenges of the consumers:

- **Security and Regulations**

Consumers may have business-critical data, which calls for protection and continuous monitoring of its access. With the cloud, the consumer may lose control of the sensitive data—for example, the consumer may not know in which country the data is being stored—and may **violate** some national data protection statutes (EU Data Protection Directive and U.S. Safe Harbor program[①]). Many regulations impose restrictions to distribute data outside of the organization's **territory**.

- **Network Latency**

Consumers may access cloud services from anywhere in the world. Although cloud resources are distributed, the resources may not be close to the consumer location, resulting in high network latency. A high network latency results in application **timeout**, thereby disabling end users from accessing the application.

- **Supportability**

Cloud may not support all applications. For example, a

---

① 欧盟数据保护指令和美国安全港计划。

consumer may want to leverage the cloud platform service for their proprietary applications, but the cloud provider may not have a compatible Operating System(OS). Also, **legacy** applications may not be supported on cloud.

- **Interoperability**

Lack of interoperability between the APIs of different cloud service providers create complexity and high migration costs for consumers when it comes to moving from one service provider to another.

The following are the challenges for the cloud service providers.

- **Service Warranty and Service Cost**

Cloud service providers usually publish a Service Level Agreement(**SLA**), so that their consumers are aware of the availability of service, quality of service, **downtime compensation**, and legal and regulatory **clauses**. Alternatively, customer-specific SLAs may be signed between a cloud service provider and a consumer. Cloud providers must ensure that they have adequate resources to provide the required level of services. SLAs typically mention penalty amount, if the cloud service providers fail to provide services. Because the cloud resources are distributed and continuously scaled to meet variable demands, it is a challenge to the cloud providers to manage physical resources and estimate the actual cost of providing the service.

- **Huge Numbers of Software to Manage**

Cloud providers, especially SaaS and PaaS providers, manage a number of applications, different Operating Systems(OSs), and middleware software to meet the needs of a wide range of consumers. This requires service providers to possess enough licenses of various software products, which, in turn, results in unpredictable **ROI**.

- **No Standard Cloud Access Interface**

Cloud service providers usually offer proprietary applications to access their cloud. However, consumers might want open APIs

---

legacy
/ˈlegəsi/
adj. 旧的;老式的
n. 遗产

warranty
/ˈwɒrənti/
n. 担保
SLA
服务级别协议
downtime
/ˈdaʊntaɪm/
n. 宕机;停工期
compensation
/ˌkɒmpenˈseɪʃn/
n. 补偿;赔偿
clause
/klɔːz/
n. 条款

ROI
/rwɑː/
n. 投资回报率

or standard APIs to become tenants of multiple clouds. This is a challenge for cloud providers because this requires an agreement among cloud providers and an upgrade of their proprietary applications to meet the standard.

Exercises

I. Fill in the blanks with the information given in the text.

1. Cloud services are accessed via the network, usually the _____, from a broad range of client platforms.

2. The physical infrastructure consists of physical IT resources that include physical servers, _____, and physical network components.

3. Cloud service providers may use physical IT resources from one or more _____ to provide services.

4. Applications and platform software are hosted on _____ to create SaaS and PaaS.

5. Cloud infrastructure management and service creation tools are responsible for managing physical and _____ infrastructures.

6. Cloud service models can be classified into three categories: IaaS, _____ and SaaS.

7. The cloud infrastructure such as servers, routers, storage, and other networking components are provided by the _____ provider.

8. _____ is the top most layer of the cloud computing stack, which is directly consumed by the end user.

9. Cloud computing can be classified into three deployment models: private, public, and _____.

10. In a _____ cloud, the services are accessible to everyone via standard Internet.

11. In a private cloud, the cloud infrastructure is operated solely for one _____ and is not shared with other organizations.

12. A high network latency results in application _____, thereby disabling end users from accessing the application.

13. Cloud service providers usually publish a _____, so that their consumers are aware of the availability of service, quality of service, downtime compensation, and legal and regulatory clauses.

II. Translate the following terms or phrases from English into Chinese.

cloud computing   resource pool

grid computing	utility computing
virtualization	service oriented architecture(SOA)
IT	CPU time
bandwidth	storage capacity
IP network	FC SAN
IP SAN	FCoE network
VLAN ID	VSAN ID
MAC address	VM
NIC	Software-as-a-Service(SaaS)
Platform-as-a-Service(PaaS)	Infrastructure as a Service(IaaS)
cloud stack	IDE
editor	compiler
builder	middleware
API	thin client
CRM	instant messaging(IM)
back-end	public cloud
private cloud	hybrid cloud
timeout	downtime

Ⅲ. Translate the following passage from English into Chinese.

Security intelligence keep the cloud safe

Many of the same characteristics that make cloud computing attractive—scalability, flexibility and accessibility, rapid application deployment, user self-service, enhanced collaboration, and location independence—can also make it challenging to secure. When a cloud is up and running today and gone tomorrow, or when the number of users quickly changes from 50 people in one city to 50,000 across the globe—how do you keep it secure? The answer lies beyond conventional security solutions, which by themselves provide neither the visibility nor the analytic capabilities necessary to proactively protect cloud environments. What today's organizations need for their clouds are integrated, comprehensive solutions that can deliver security intelligence. Advanced security intelligence solutions can close security gaps by using labor-saving automation to analyze millions of events occurring within the cloud, and discover system vulnerabilities through the normalization and correlation of these events. These tools can then remove false positives, add context and reduce vulnerabilities to a smaller, more manageable number for security teams to act upon.

Section B   The Internet of Things

Ⅰ. What is the Internet of Things

It is foreseeable that any object will have a unique way of identification in the coming future, what is commonly known in the networking field of computer sciences as Unique Address, creating an **addressable continuum** of computers, sensors, **actuators**, mobile phones; i.e. any thing or object around us. Having the capacity of addressing each other and verifying their identities, all these objects will be able to exchange information and, if necessary, actively process information according to predefined schemes, which may or may not be **deterministic**.

The definition of "Internet of Things" has still some fuzziness, and can have different **facets** depending on the perspective taken. Considering the functionality and identity as central it is reasonable to define the IoT as "Things having identities and virtual personalities operating in smart spaces using intelligent interfaces to connect and communicate within social, environmental, and user contexts". A different definition, that puts the focus on the seamless integration, could be formulated as "Interconnected objects having an active role in what might be called the Future Internet".

II. Technologies for the Internet of Things

The Internet of Things is a technological revolution that represents the future of computing and communications, and its development depends on dynamic technical innovation in a number of important fields, from wireless **sensors** to **nanotechnology**.

Firstly, in order to connect everyday objects and devices to large databases and networks—and indeed to the network of networks(the internet)—a simple, **unobtrusive** and cost-effective system of item identification is crucial. Only then can data about things be collected and processed. Radio-frequency identification (**RFID**) offers this functionality.

Secondly, data collection will benefit from the ability to detect changes in the physical status of things, using sensor technologies. **Embedded** intelligence in the things themselves can further enhance the power of the network by devolving information processing capabilities to the edges of the network.

Finally, advances in **miniaturization** and nanotechnology mean that smaller and smaller things will have the ability to interact and connect. A combination of all of these developments will create an Internet of Things that connects the world's objects in both a sensory and an intelligent manner.

- **RFID**

RFID(see Figure 7B-1) technology, which uses radio waves to identify items, is seen as one of the **pivotal** enablers of the Internet of Things. Although it has sometimes been **labelled** as the next-generation of **bar codes**, RFID systems offer much more in that they can track items in real-time to yield important information about their location and status. Early applications of RFID include automatic highway toll collection, supply-chain management(for

Figure 7B-1 The Composition of RFID

large retailers), **pharmaceuticals**(for the prevention of counterfeiting) and e-health(for patient monitoring). More recent applications range from sports and **leisure**(ski passes) to personal security (tagging children at schools). RFID tags are even being implanted under human skin for medical purposes, but also for VIP access to bars. E-government applications such as RFID in drivers' licences, passports or cash are under consideration. RFID readers are now being embedded in mobile phones. Nokia, for instance, released its RFID-enabled phones for businesses with workforces in the field in mid-2004.

- **Sensors**

In addition to RFID, the ability to detect changes in the physical status of things is also essential for recording changes in the environment. In this regard, sensors(see Figure 7B-2) play a pivotal role in bridging the gap between the physical and virtual worlds, and enabling things to respond to changes in their physical environment. Sensors collect data from their environment, generating information and raising awareness about context. For example, sensors in an electronic jacket can collect information about changes in external temperature and the parameters of the jacket can be adjusted accordingly.

Figure 7B-2 Some Examples of Sensors

- **Embedded Intelligence**

Embedded intelligence in things themselves will distribute processing power to the edges of the network, offering greater possibilities for data processing and increasing the **resilience** of the network. This will also empower things and devices at the edges of the network to take independent decisions. "Smart things" are difficult to define, but imply a certain processing power and reaction to external **stimuli**. Advances in smart homes, smart vehicles and personal robotics are some of the leading areas. Research on wearable computing (including wearable mobility vehicles) is **swiftly** progressing. Scientists are using their imagination to develop new devices and appliances, such as intelligent ovens that can be controlled through phones or the internet, online refrigerators and networked blinds(see Figure 7B-3).

Figure 7B-3   Some Embedded Intelligence

The Internet of Things will draw on the functionality offered by all of these technologies to realize the vision of a fully interactive and responsive network environment.

Ⅲ. Wider Technological Trends

It is possible to identify, for the years to come, four distinct macro-trends that will shape the future of IT, together with the explosion of **ubiquitous** devices that constitute the future Internet of Things.

(1) The first one, sometimes referred as "exaflood[①]" or "data **deluge**", is the explosion of the amount of data collected and exchanged. Just to give some numbers, business forecasts indicate that in the year 2015 more than 220 **Exabytes** of data will be stored. As current network are ill-suited for this exponential traffic growth, there is a need by all the actors to re-think current networking and storage architectures. It will be **imperative** to find novel ways and mechanisms to find, fetch, and transmit data. One relevant reason for this data deluge is the explosion in the number of devices collecting and exchanging information as envisioned as the Internet of Things becomes a reality.

(2) The energy required to operate the intelligent devices will dramatically decreased. Already today many data centres have

---

① exaflood：研究人员杜撰出的术语"数字洪水"，描绘互联网上迅猛增长的数据流量。

reached the maximum level of energy consumption and the **acquisition** of new devices has necessarily to follow the **dismissal** of old ones. Therefore, the second trend can be identified covering all devices and systems from the tiniest smart dust to the huge data centres: the search for a zero level of entropy where the device or system will have to harvest its own energy.

(3) Miniaturisation of devices is also taking place amazingly fast. The objective of a single-electron transistor is getting closer, which seems the ultimate limit, at least until new discoveries in physics.

(4) Another important trend is towards autonomic resources. The ever growing complexity of systems will be unmanageable, and will hamper the creation of new services and applications, unless the systems will show self-* properties, such as self-management, self-healing and self-configuration.

Ⅳ. Applications of the Internet of Things

The IoT is connecting new places—such as manufacturing floors, energy grids, healthcare facilities, and transportation systems—to the Internet. When an object can represent itself digitally, it can be controlled from anywhere. This connectivity means more data, gathered from more places, with more ways to increase efficiency and improve safety and security.

- **Retail**

The first large scale application of the Internet of Things technologies, will be replace the bar code in retail. The main barriers so far have been the much higher cost of the **tag** over the bar code, some needed technology improvement for what concerns transmission of metals and liquid items, and privacy concerns. **Nonetheless**, the replacement has already started in some pilot projects and although one may expect to see co-existence of the two identification mechanisms for many years into the future, advances in the electronics industry will render the RFID tag ever cheaper and more attractive and accessible to the retailers.

- **Logistics**

It is important to remember that innovation in logistics normally does not change the industry fundamentally but allows improving efficiency of processes or enables new value adding features. The first observation to be made from the preceding discussion is that the warehouses will become completely automatic with items being checked in and out and orders automatically passed to the suppliers. This will allow better asset management and **proactive** planning on transported without human intervention from producer to consumer and the manufacturers will have a direct feedback on the behalf of the transporter. Goods may be market's needs. In this way the production and transportation can be adapted dynamically thus saving time, energy, and the environment. Figure 7B-4 shows an example of electronic logistics.

Figure 7B-4　An Example of Electronic Logistics

- **Intelligent Home**

There are already examples of smart houses being demonstrated and the future **intelligent home** will build on these experiences. The present experience is **tailor** made, and each thing in the house has been carefully selected and tuned to interoperate with all the other intelligent devices. This is too costly for most houses and the intelligent home remains a dream for most people. The big paradigm shift comes when every smart object knows the interoperable protocols removing the need for the dedicated systems developed independently today. For instance, there are several solutions for intelligently controlling every **power socket** in the house thus allowing simple tasks like switching on and off lights, and more complex ones such as **fine-grained** management of **electrical heaters**, in order to set the **ambient** temperature. However, the control systems in operation today are quite basic and apply only to the wall socket, and can not manage appliances connected through extension cords. In the future Internet of Things the lamps or even the light bulbs will be addressable and intelligent, and a global house management controller will be able to control every single smart device(see Figure 7B-5).

Figure 7B-5　Intelligent Home

Exercises

I. Fill in the blanks with the information given in the text.

1. Having the capacity of _____ each other and verifying their identities, all these objects will be able to exchange information.

2. Things having _____ and virtual personalities operating in smart spaces using intelligent interfaces to connect and communicate within social, environmental, and user contexts.

3. Data collection will benefit from the ability to detect changes in the physical status of things, using _____ technologies.

4. Advances in _____ and nanotechnology mean that smaller and smaller things will have the ability to interact and connect.

5. Research on _____ computing(including wearable mobility vehicles) is swiftly progressing.

6. The _____ required to operate the intelligent devices will dramatically decreased.

7. When an object can represent itself _____, it can be controlled from anywhere.

II. Translate the following terms or phrases from English into Chinese.

Internet of Things(IoT)　　　　wireless sensor
RFID　　　　　　　　　　　　radio wave
bar code　　　　　　　　　　data deluge
smart house

Section C Supercomputer

Ⅰ. Supercomputer

Supercomputers[①] are capable of computing and processing data. They are characterized by high speed and large **capacity**. They are equipped with a variety of external and **peripheral** devices and rich and high-function software systems.

Supercomputers are designed in a **turbo**-like fashion, with each **blade** acting as a server that can work together and be added or subtracted as the application requires. Generally speaking, supercomputers can perform more than 10 million calculations per second and store more than 10 million bits.

The development of supercomputer is an important development direction of electronic computer. Its development level marks the degree of a country's science and technology and industrial development, reflecting the strength of the country's economic development.

Ⅱ. Supercomputers in China

In an increasingly **fierce uphill** race of **exascale** computing worldwide, more Chinese supercomputers are making the world's

capacity
/kəˈpæsəti/
n. 容量
peripheral
/pəˈrɪfərəl/
adj. 外围的
turbo
/ˈtɜːbəʊ/
n. 涡轮
blade
/bleɪd/
n. 刀片
fierce
/fɪəs/
adj. 激烈的
uphill
/ˌʌpˈhɪl/
adj. 向上的
exascale
/ˌɪɡˈzæskeɪl/
百亿亿次级

---

① supercomputer：超级计算机，也称为超算。

Top 500 list as seen in the latest edition published at ISC① High Performance.

In 2009, China's National University of Defense Technology② unveiled the Tianhe-1③ supercomputer with a **peak performance** of 1.206 **petaflops** per second. Tianhe-1, China's first **gigabit** supercomputer, is deployed at the National Supercomputer Center in Tianjin④. The Tianhe-1 held the top spot on the edition of the Top 500 supercomputer list, in November 2010. China has become the second country after the United States to independently develop a petaflops supercomputer.

In particular, the emergence of Sunway Taihulight⑤ in 2016 marked China's entry into the world's leading position in supercomputing. The Sunway Taihulight supercomputer was developed by the National Parallel Computer Engineering and Technology Research Center⑥ and installed in the Wuxi National Supercomputing Center⑦. The Sunway Taihulight supercomputer is equipped with 40,960 China-developed Sunway 26010 **multi-core** processors, which use the 64-bit **autonomous** Sunway instruction system and have a peak performance of 3,168 **teraflops**, a core operating frequency of 1.5GHz. Sunway Taihulight supercomputer ranked first on the Top 500 list in 2016 and 2017 for its relatively high computing speed.

peak
/piːk/
n. 峰值
peak performance
峰值性能
petaflop
/ˈpiːtəflɒp/
n. 每秒千万亿次浮点运算
gigabit
/ˈɡɪɡəbɪt/
n. 千兆位
autonomous
/ɔːˈtɒnəməs/
adj. 自主的
teraflop
/ˈterəflɒp/
n. 每秒万亿次浮点运算

---

① ISC：International Supercomputer Conference，国际超级计算机大会。1986 年，已故的汉斯·维尔纳·梅尔在曼海姆大学组织了世界上第一个"超级计算机研讨会"，时至今日，它已成为世界上历史最悠久、欧洲范围内最重要的超级计算大会。
② National University of Defense Technology：国防科学技术大学，位于湖南省长沙市，取得了以"天河"系列超级计算机系统、"北斗"卫星导航定位系统关键技术、磁浮列车等为代表的一大批自主创新成果。
③ Tianhe-1："天河一号"超级计算机，中国首台千兆次超级计算机，峰值性能为每秒 1.206 千万亿次，2010 年 11 月高居超级计算机 500 强榜首。
④ National Supercomputer Center in Tianjin：位于天津的国家超级计算机中心。
⑤ Sunway Taihulight："神威·太湖之光"超级计算机，安装了 40960 个中国自主研发的神威 26010 众核处理器，采用 64 位自主神威指令系统，峰值性能达到 3168 万亿次每秒，核心工作频率为 1.5GHz。2016 年、2017 年连续两年蝉联全球超级计算机 500 强冠军。
⑥ National Parallel Computer Engineering and Technology Research Center：国家并行计算机工程技术研究中心，1992 年 8 月组建，中心总部设在北京市海淀区。
⑦ Wuxi National Supercomputing Center：国家超级计算无锡中心。

Ⅲ. The Top 5 in 2021

Japanese supercomputer "Fugaku①" held the top spot on the new edition of the Top 500 supercomputer list, in July, 2021, while China continues to lead in the number of Top 500 supercomputers. China continues to dominate the list with regard to system number, claiming 188 supercomputers on the list. The United States is number two with 122 systems, and Japan is third with 34.

- Fugaku has a High Performance Linpack② result of 442 petaflops, besting the second-placed "Summit③" system of the United States by three times. Fugaku's peak performance is above an **exaflop**. Such an achievement has caused some to introduce this machine as the first "Exascale" supercomputer④.
- The "Summit" is an IBM-built supercomputer running at Oak Ridge National Laboratory⑤ in the U.S. state of **Tennessee**. It remains the country's fastest supercomputer.
- At number three is "Sierra⑥", a system at the Lawrence Livermore National Laboratory⑦ in the U.S. state of **California**.
- Chinese supercomputers "Sunway TaihuLight" ranks fourth on the Top500 list. Chinese manufacturers dominate the list in the number of installations, with **Lenovo**⑧ and **Inspur**⑨ claiming top two, and **Sugon**⑩ standing at the fourth.

exaflop
/ˌɪgˈzæflɒps/
n. 百亿亿次
Tennessee
/ˌtenəˈsiː/
n. (美国)田纳西州
California
/ˌkæləˈfɔːniə/
n. 加利福尼亚
Lenovo
/lɪˈnəʊvəʊ/
n. 联想(公司名)
Inspur
/inˈspə/
n. 浪潮集团
Sugon
/ˈsugən/
n. 中科曙光

---

① Fugaku：富岳，日本开发的超级计算机，2021年高居超级计算机500强榜首，峰值浮点性能高达513千万亿次每秒。
② Linpack：国际上使用最广泛的测试高性能计算机系统浮点性能的基准测试。
③ Summit：美国最强大的超级计算机，由IBM公司制造，于2018年诞生，取代了我国"神威·太湖之光"世界超级计算机第一的位置。2020年被日本的Fugaku取代，位居世界超级计算机第二位。
④ "Exascale" supercomputer：百亿亿次超级计算机。
⑤ Oak Ridge National Laboratory：美国田纳西州橡树岭国家实验室。
⑥ Sierra：美国的超级计算机，2021年世界超级计算机500强排名第三。
⑦ Lawrence Livermore National Laboratory：劳伦斯-利弗莫尔国家实验室。
⑧ Lenovo：联想集团，1984年创办，首任CEO为柳传志，最初的英文名是Legend，2003年启用新标识Lenovo。联想是目前全球最大的顶级超算供应商，在2021年全球性能最强的500台高性能计算集群中，联想集团交付了184台。
⑨ Inspur：浪潮集团，中国本土顶尖的大型IT企业之一。
⑩ Sugon：中科曙光，国内高性能计算领域的领军企业。

Supercomputers installed by the three Chinese vendors account for 281 of the Top500 systems.
- At number five is the "Perlmutter①" system at the U.S. Lawrence Berkeley National Laboratory②, which is the only new entry in the Top10.

IV. Application in COVID-19③

Supercomputers are essential **infrastructure** for modern technological innovation. It is able to provide support for scientific research, development of artificial intelligence and cloud computing. More than 30 percent of technological innovations in developed countries such as the US, Europe and Japan rely on supercomputing.

In this early April, 2020, the AI-supported assistance system to analyze CT④ images of COVID-19 patients opened to the world, it has benefited more than 100 overseas medical institutions. Total visits during peak hours reached more than 2000 per day.

The National Supercomputer Center in Tianjin jointly tested and adopted this system with hundreds of experts from over 50 research and development institutions worldwide, and from over 30 hospitals across the nation.

The National Supercomputer Center in Tianjin owns the Tianhe-1 supercomputer and Tianhe-3 supercomputer prototype system. Now the center provides high-end information technology service, such as high-performance computing, cloud computing and big data, for **nationwide** research institutes, universities and key enterprises.

The Tianhe-1 supercomputer works on more than 8000 scientific research computing projects at full load every day.

To develop economy and industry service, over 10 specialized

infrastructure
/ˈɪnfrəstrʌktʃə/
n. 基础设施

nationwide
/ˌneɪʃnˈwaɪd/
adj. 全国性的

---

① Perlmutter：美国的超级计算机，2021年世界超级计算机500强排名第五。
② Lawrence Berkeley National Laboratory：劳伦斯伯克利国家实验室。
③ COVID-19：Corona Virus Disease 2019，2019年爆发的全球大流行的新型冠状病毒肺炎。
④ CT：Computed Tomography，电子计算机断层扫描。CT是用X射线束对人体某部位一定厚度的层面进行扫描，由探测器接收透过该层面的X射线，转变为可见光后，由光电转换变为电信号，再经模拟/数字转换器转为数字，输入计算机处理，用于探测人体疾病。

platforms in many domains have been built relying Tianhe-1, such as oil exploration, advanced material and smart city.

Exercises

Ⅰ. Fill in the blanks with the information given in the text.

1. _____ are characterized by high speed and large capacity. They can perform more than 10 million calculations per second and store more than 10 million bits.

2. China's National University of Defense Technology unveiled the _____ supercomputer in 2009.

3. Sunway _____ supercomputer ranked first on the Top 500 list in 2016 and _____ for its relatively high computing speed.

4. Japanese supercomputer Fugaku's peak performance is above an exaflop. Such an achievement has caused some to introduce this machine as the first _____ supercomputer.

5. In 2021, China continues to dominate the Top 500 with regard to system _____, claiming 188 supercomputers on the list.

Ⅱ. Translate the following terms or phrases from English into Chinese.

supercomputer	large capacity
a turbo-like fashion	peak performance
National Supercomputer Center	China-developed
instruction system	multi-core

Unit Eight Artificial Intelligence

Section A Artificial Intelligence

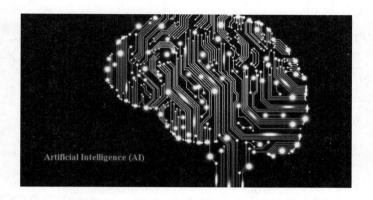

Ⅰ. What is AI?

In 1955, John McCarthy①, one of the pioneers of AI, was the first to define the term **artificial intelligence**, roughly as follows: the goal of AI is to develop machines that behave as though they were intelligent.

In the book *Artificial Intelligence, A New Synthesis*, the author Nils J. Nilsson② said, Artificial Intelligence (AI) broadly defined, is concerned with intelligent behavior in **artifacts**, intelligent behavior, in turn, involves **perception**, reasoning, learning, communicating and acting in complex environments. AI has as one of its long-term goals the development of machines that can do these things as well as humans can, or possibly even better. Another goal of AI is to understand this kind of behavior whether it occurs in machines or in humans or other animals, Thus, AI has both engineering and scientific goals.

**artificial
intelligence(AI)**
人工智能
synthesis
/ˈsɪnθəsɪs/
n. 综合
artifact
/ˈɑːtɪfækt/
n. 人工制品
perception
/pəˈsepʃn/
n. 感知

---

① John McCarthy：约翰·麦卡锡(1927—2011)，美国计算机科学家与认知科学家，他首次提出 Artificial Intelligence 一词，被称为"人工智能"之父。
② Nils J. Nilsson：尼尔斯·约翰·尼尔森(1933—2019)，人工智能领域的开创者之一。A * 搜索算法发明人。

AI as a science in its own right has only existed since the middle of the Twentieth Century. AI is interdisciplinary, for it draws upon interesting discoveries from such diverse fields as logic, operations research, **statistics**, control engineering, image processing, **linguistics**, **philosophy**, **psychology**, and **neurobiology**. On top of that, there is the subject area of the particular application.

statistics
/stəˈtɪstɪks/
n. 统计学
linguistics
/lɪŋˈɡwɪstɪks/
n. 语言学
philosophy
/fəˈlɒsəfi/
n. 哲学
psychology
/saɪˈkɒlədʒi/
n. 心理学
neurobiology
/ˌnjʊərəʊbaɪˈɒlədʒi/
n. 神经生物学

Ⅱ. The History of AI

The history of the various AI areas is shown in Figure 8A-1. The width of the bars indicates pre valence of the method's use.

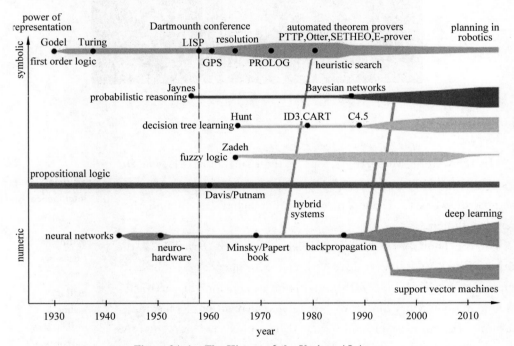

Figure 8A-1 The History of the Various AI Areas

- **The First Beginnings**

In the 1930s, Kurt Godel[①], Alonso Church[②], and Alan

---

① Kurt Godel：库尔特·哥德尔(1906—1978)，美籍奥地利数学家、逻辑学家和哲学家，是20世纪最伟大的逻辑学家之一，其最杰出的贡献是哥德尔不完全性定理。
② Alonso Church：阿隆索·丘奇，逻辑学家，1944年出版著作 *Introduction to Mathematical Logic*。

Turing laid important foundations for logic and **theoretical** computer science.

In the 1940s, based on results from **neuroscience**, McCulloch①, Pitts② designed the first mathematical models of neural networks MP Model(see Figure 8A-2). However, computers at that time lacked sufficient power to **simulate** simple brains.

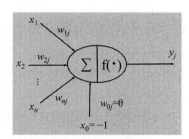

Figure 8A-2　MP Model

- **AI Rises**

In 1950, Alan Turing made a name for himself as an early pioneer of AI with his definition of an intelligent machine with **the Turing test** and wrote about learning machines and **genetic** algorithms.

In 1951, Marvin Minsky③ developed a neural network machine. With 3000 vacuum bubes he simulated 40 neurons.

In 1956, McCarthy organized a historic conference in Dartmouth College. Here the name Artificial Intelligence was first introduced.

- **Logic Solves(Almost) All Problems**

AI as a practical science of thought mechanization could of course only begin once there were programmable computers.

---

① McCulloch：Warren S.McCulloch,神经生理学家,1943 年,他与 Pitts 建立了神经网络和数学模型,称为 MP 模型,从而开创了人工神经网络研究的时代。
② Pitts：Walter Pittsm,沃尔特·皮茨(1923—1969),数理逻辑学家。
③ Marvin Minsky：马文·明斯基(1927—2016),"人工智能"之父和"框架理论"的创立者,1951 年创造了世界上第一台具有学习能力的机器。

In the 1950s, Newell① and Simon② introduced Logic Theorist, the first **automatic theorem prover**, and thus also showed that with computers, which actually only work with numbers, one can also process symbols. At the same time McCarthy introduced the language LISP a programming language specially created for the processing of symbolic structures.

In the 1970s, the logic programming language PROLOG③ was introduced as the European counterpart to LISP④.

Until well into the 1980s, a breakthrough spirit dominated AI, especially among many **logicians**. Japan began, at great expense, the "Fifth Generation Project" with the goal of building a powerful PROLOG machine.

Computers, physicists were able to show, using computers which were now sufficiently powerful, that mathematically modeled neural networks were capable of learning using **training examples**, to perform tasks which previously required costly programming, especially in **pattern recognition**. Facial recognition in photos and handwriting recognition were two example applications.

- **Reasoning Under Uncertainty**

But soon even feasibility limits became obvious. The neural networks could acquire impressive capabilities, but it was usually not possible to capture the learned concept in simple **formulas** or logical rules. Attemplts to combine neural nets with logical rules or the knowledge of human experts met with great difficulties. Additionally, no satisfactory solution to the structuring and modularization of the networks was found.

AI as a practical, goal-driven science searched for a way out of

---

① Newell：Allen Newell,艾伦·纽厄尔(1927—1992),美国计算机科学和认知信息学领域的科学家。
② Simon：Herber A. Simon,艾尔伯特·西蒙(1916—2001),美国计算机领域著名的科学家,他和 Newell 一起成功开发了世界上最早的启发式程序"逻辑理论家",开创了机器定理证明这一新的学科领域。
③ PROLOG：Programming in Logic,一种面向演绎推理的逻辑型程序设计语言。应用于许多领域,如关系数据库、数理逻辑、抽象问题求解、自然语言理解和专家系统等。
④ LISP：List Processing,适用于符号处理、自动推理、硬件描述和超大规模集成电路设计等,是最有影响、使用广泛的人工智能语言。

---

automatic theorem prover
自动定理证明

logician
/ləˈdʒɪʃn/
n. 逻辑学家

training examples
训练样本
pattern recognition
模式识别

Reasoning Under Uncertainty
不确定性推理

formula
/ˈfɔːmjələ/
n. 公式

this crisis. The most promising, **probabilistic** reasoning, worked with **conditional probabilities** for **propositional calculus** formulas. In 1990, Pearl[①], etc., brought probability theory into AI with Bayesian networks[②]. The success of Bayesian networks stemmed from their intuitive comprehensibility, the clean semantics of conditional probability, and from the centuries-old, mathematically grounded probability theory.

The weaknesses of logic, which can only work with two truth values, can be solved by **fuzzy logic**, which pragmatically introduces **infinitely** many values between zero and one. It is successfully utilized, especially in control engineering.

A much different path led to the successful **synthesis** of logic and neural networks under the name **hybrid** systems. For example, neural networks were employed to learn **heuristics** for reduction of the huge combinatorial search space in proof discovery.

Methods of **decision tree**[③] learning from data also work with probabilities. System like CART[④], ID3[⑤] and C4.5[⑥] can quickly and automatically build very accurate decision trees which can represent propositioanl concepts and then be used as **expert systems**. Today they are a favorite among machine learning techniques.

Since about 1990, **data mining** has developed as a subdiscipline of AI in the area of statistical data analysis for extraction of knowledge from large database.

From statistical learning theory, Vapnik[⑦] developed support vector machines[⑧], in 1995, which are very important today.

---

① Pearl：Judea Pearl，美国以色列裔计算机科学家和哲学家，"贝叶斯网络"之父，他将概率和因果推理的算法引入人工智能领域，取得杰出的成就。
② Bayesian networks：贝叶斯网络，是目前不确定知识表达和推理领域中最有效的理论模型之一，已成功用于医疗诊断、统计决策、专家系统、学习预测等领域。
③ decision tree：决策树，在机器学习中是一个预测模型。
④ CART：Classification And Regression Tree，分类与回归树，一种非常有趣且十分有效的非参数分类和回归方法，通过构建二叉树达到预测目的。
⑤ ID3：Iterative Dichotomizer 3，一种贪心算法，用来构造决策树。以信息论为基础，以信息熵和信息增益度为衡量标准，实现对数据的归纳分类。
⑥ C4.5：用于产生决策树的算法，是对 ID3 算法的一个扩展。
⑦ Vapnik：Vladimir Naumovich Vapnik，弗拉基米尔·万普尼克，俄罗斯统计学家、数学家。他是统计学习理论的主要创建人之一。
⑧ support vector machines(SVM)：支持向量机。1964 年提出，20 世纪 90 年代后期得到快速发展，并衍生出一系列改进和扩展算法，在人像识别、文本分类等模式识别中得到应用。

---

probabilistic
/ˌprɒbəˈlɪstɪk/
adj. 概率的
conditional probabilities
条件概率
propositional
/ˌprɒpəˈzɪʃənl/
adj. 命题的
calculus
/ˈkælkjələs/
n. 微积分；演算
fuzzy logic
模糊逻辑
infinitely
/ˈɪnfɪnətli/
adv. 无限地
synthesis
/ˈsɪnθəsɪs/
n. 综合
hybrid
/ˈhaɪbrɪd/
n. 混合
heuristics
/hjuˈrɪstɪks/
n. 启发式
expert system
专家系统
data mining
数据挖掘

- **Distributed, Autonomous and Learning Agents**

Distributed artificial intelligence(DAI), has been an active area research since about 1985. One of its goals is the use of parallel computers to increase the efficiency of problem solvers.

A very different conceptual approach results from the development of autonomous software agents and robots that are meant to cooperate like human teams. Only the cooperation of many agents leads to the intelligent behavior or to the solution of a problem. Multi-agent systems become popular.

Active skill **acquisition** by robots is an exciting area of current research. There are robots today, for example, that independently learn to walk or to perform various motor skills related to soccer.

On May 11, 1997, an IBM computer called IBM Deep Blue[①] beat the world chess champion after a six-game match: two wins for IBM, one for the champion and three **draws**. The match lasted several days. Behind the contest, however, was important computer science, pushing forward the ability of computers to handle the kinds of complex calculations needed to help discover new medical drugs; do the broad financial modeling needed to identify trends and do risk analysis; handle large database searches; and perform massive calculations needed in many fields of science.

In 2006, Service robotics became a major AI research area.

In 2009, First Google self-driving car drove on the California freeway.

- **AI Revolution**

Around the year 2010 after 25 years of research on neural networks, scientists could start harvesting the fruits of their research. The very powerful Deep Learning[②] networks can for

acquisition
/ˌækwɪˈzɪʃn/
n. 获得

draw
/drɔː/
n. 平局

---

① Deep Blue：深蓝是美国 IBM 公司生产的一台超级国际象棋计算机,采用并行计算,有 32 个微处理器,1997 年版本的深蓝每秒可以计算 2 亿步棋,输入了一百多年来优秀棋手的对局两百多万局。1997 年 5 月 11 日,计算机首次击败了排名世界第一的棋手加里·卡斯帕罗夫,机器的胜利标志着国际象棋历史的新时代。

② Deep Learning(DL)：深度学习,机器学习领域中一个新的研究方向,在搜索技术、数据挖掘、机器翻译、自然语言处理、多媒体学习、语音、推荐和个性化技术等领域取得了很多成果。深度学习使机器模仿视听和思考等人类活动,解决了很多复杂的模式识别难题,使人工智能相关技术取得了很大进步。

example absorb huge amounts of **unstructured data** such as text, images, and audio, can learn to classify images with very high arruracy. Since image classification is of crucial importance for all types of smart robots, this **initiated** the AI revolution which in turn leads to smart self-driving cars and service robots.

Table 8A-1　Major Achievements in AI Fields in 2015-2016

2015	**Daimler** premiers the first autonomous truck on the **Autobahn**
	Google self-driving cars have driven over one million miles within cities
	Deep Learning enables very good image classification
	Painting in the style of the Old Masters can ba automatically generated with deep learning. AI becomes creative
2016	AlphaGo by Google DeepMind beats the world's best **Go** player

AlphaGo①, an artificial intelligence program, featuring "deep learning algorithm", developed by Google Deep Mind, scored a 4∶1 victory against professional champion Lee Sedol② of the Republic of Korea(see Figure 8A-3) in 2016, and a 3∶0 match with the world's number one player Ke Jie③ in China in May 2017. AlphaGo is almost unbeatable in Go because of its fast computing ability, which allows it to review many moves and manuals within a very short time. Essentially, AlphaGo is an imitation of the human brain in terms of learning ability.

Figure 8A-3　Lee Sedol and AlphaGo

AlphaGo symbolizes that computer technology has entered the new information technology era of artificial intelligence, which is

---

① AlphaGo：阿尔法围棋,由谷歌旗下 DeepMind 公司开发的人工智能机器人。AlphaGo 用到了很多新技术,如神经网络、深度学习、蒙特卡洛树搜索法等,使其实力有了实质性飞跃。2016 年 3 月,它以 4∶1 战胜围棋世界冠军、职业九段棋手李世石,2017 年 5 月,在中国乌镇以 3∶0 战胜排名世界第一的世界围棋冠军柯洁。
② Lee Sedol：李世石,1983 年出生,韩国职业围棋手,世界顶级围棋棋手。
③ Ke Jie：柯洁,1997 年出生,2014 年 8 月,他在世界围棋等级分中首次排名第一。2017 年 5 月与 AlphaGo 进行人机大战,虽 3 局均败,但次局被机器评定表现完美。

characterized by big data, big computing, and big decision-making. It is approaching humans in intelligence.

Ⅲ. Prospect

In the modern world, we are surrounded by AI. From assistants such as Amazon's Alexa①, smart TV, wearable device, smart robot to the internet predicting what we may like to buy next, AI is found everywhere. Also self-driving cars are an example of the application of AI. But AIs are still incompetent when it comes to creative thinking.

Artificial General Intelligence(**AGI**)② till date remains just a concept. The idea behind AGI is to make it as adaptable and flexible as human intelligence. When scientists will be able to develop AGI remains a hotly contested debate, with some saying it'll arrive by as soon as 2040 to others saying its centuries away, given the lack of understanding of the human brain.

Exercises

Ⅰ. Fill in the blanks with the information given in the text.

1. In 1955, John McCarthy was the first to define the term artificial intelligence: the goal of AI is to develop machines that _____ as though they were _____.

2. Nils J. Nilsson said, Artificial Intelligence broadly defined, is concerned with intelligent behavior in artifacts, intelligent behavior, in turn, involves, _____, reasoning, _____, communicating and acting in complex environments.

3. AI is interdisciplinary, for it draws upon interesting discoveries from such diverse fields as _____, operations research, _____, control engineering, image processing, linguistics, philosophy, psychology, and _____.

4. In the 1940s, based on results from neuroscience, McCulloch, Pitts designed the first mathematical models of neural networks _____ Model.

5. From statistical learning theory, Vapnik developed _____, in 1995, which are very important today.

---

① Alexa：亚马逊的一家子公司，每天在网上搜集超过 1000GB 的信息，是当前拥有 URL 数量最庞大，排名信息发布最详尽的网站。
② AGI：artificial general intelligence，通用人工智能，这一领域主要专注于研制像人一样思考、像人一样从事多种用途的机器。

6. On May 11, 1997, an IBM computer called _____ beat the world chess champion, pushing forward the ability of computers to handle the kinds of complex calculations.

7. AlphaGo, an artificial intelligence _____, is almost unbeatable in Go, featuring "_____ algorithm".

II. Translate the following terms or phrases from English into Chinese.

artificial intelligence	control engineering
image processing	neural network
Turing test	genetic algorithm
training example	pattern recognition
conditional probabilities	Bayesian network
fuzzy logic	decision tree
expert system	data mining
SVM	self-driving
DL	

III. Translate the following passage from English into Chinese.

Turing test

Alan Turing define an intelligent machine, in which the machine in question must pass the following test. The test person Alice sits in a locked room with two computer terminals. One terminal is connected to a machine, the other with a non-malicious person Bob. Alice can type questions into both terminals. She is given the task of deciding, after five minutes, which terminal belongs to the machine. The machine passes the test if it can trick Alice at least 30% of the time.

Section B Metaverse

Ⅰ. Introduction

The word, **metaverse**, was coined in Neal Stephenson's 1992 science fiction movel "Snow Crash①", where humans, as **avatars**, interact with each other and software agents, in a three-dimensional space that uses the **metaphor** of the real world. The book's main character, named Hiro Protagonist②, delivers pizza for the **Mafia**. When not working, Mr Protagonist plugs into the Metaverse: a networked virtual reality in which people appear as self-designed "avatars" and engage in activities both **mundane** (conversation) and **extraordinary**(sword fights).

The Metaverse is a collective virtual shared space, created by the convergence of virtually enhanced physical reality and physically **persistent** virtual space, including the sum of all virtual worlds, augmented reality③, and the internet. The word metaverse combines the prefix "meta"(meaning "beyond") with "universe" and is typically used to describe the concept of a future iteration of the internet, made up of persistent, shared, decentrilized, 3D virtual spaces linked into a perceived virtual universe.

A metaverse environment might look a lot like many of the online video games that are popular today. These involve players

metaverse
/ˈmetəvɜːs/
n. 元宇宙；虚拟世界
avatar
/ˈævətɑː/
n. 虚拟人物
metaphor
/ˈmetəfə/
n. 隐喻；比喻
Mafia
/ˈmæfiə/
n. 黑手党
mundane
/mʌnˈdeɪn/
n. 平凡的事
extraordinary
/ɪkˈstrɔːdnri/
n. 非凡的事
beyond
/bɪˈjɒnd/
prep. 超越
persistent
/pəˈsɪstənt/
adj. 持久的

---

① Snow Crash：《雪崩》，美国作家尼尔·斯蒂芬森于 1992 年出版的一部科幻小说，书中第一次提出了 metaverse 这个词。
② Hiro Protagonist：《雪崩》小说中的主人公名字，小宏（日本名）。
③ augmented reality：缩写为 AR，即增强现实。

around the world interacting in virtual environments and even permit users to buy digital items with real money.

II. Related Technologies

Metaverse describes a non-physical world in which individuals can interact through different kinds of virtual technology.

For example, Figure 8B-1 shows that a metaverse could permit people living on different sides of the world to meet up through technology and virtually go on a vacation, play sports or work together on projects. People linked to the metaverse would be connected at all times and physical distance would not limit their ability to interact.

Figure 8B-1　A Virtual Event

The main technologies that would drive such a world would be virtual reality(VR①), augmented reality(AR), AI technology, digital people(virtual **idols**), cloud **rendering**, etc. Other, yet-to-be invented technologies would likely also be used to improve experiences within the metaverse.

- **VR**

Virtual reality, referred to as VR technology, also known as artificial environment. VR technology will bring the **audio-visual** experience to a new height.

The use of computers or other intelligent computing devices to

idol
/ˈaɪdl/
n. 幻象；偶像

rendering
/ˈrendərɪŋ/
n. 渲染

audio-visual
视听

---

① virtual reality：VR,虚拟现实,一种可以创建和体验虚拟世界的计算机仿真系统,使用沉浸到该环境中。虚拟现实具有一切人类所拥有的感知功能,比如听觉、视觉、触觉、味觉、嗅觉等,实现人机交互。

simulate a virtual world of three degrees, providing users with visual, auditory, **tactile** and other sensory simulation, so that the user as **immersive**.

Core value is below.

(1) Evolution of display mode: the traditional **plane display** mode is upgraded to a **panoramic** display, which greatly increases the user's sense of immersion and the degree of simulation of the content.

(2) Horizontal positioning **deception**: the visual angle of the user is simulated by horizontal positioning system, and the vision of the high quality image is displayed.

(3) 3D sound control hearing: the use of the most advanced 3D sound solutions to simulate surround auditory experience, allowing users to feel immersive.

(4) A variety of interactive ways: combined with **handle manipulation**, behavior detection, speech recognition and other interactive ways to improve the user's behavior even tactile interaction experience.

- **AR**

Augmented reality (AR[①]) is a new technology developed on the basis of virtual reality, also known as mixed reality. It is provided by the computer system information technology increasing the user's perception of the real world, the application of the virtual information into the real world and computer-generated virtual objects, scene of system information **superimposed** on the real scene, so as to realize the enhancement of reality. The example of AR is shown in Figure 8B-2.

A perfect AR system includes a number of disciplines, including the technology of system display and positioning technology, virtual reality **fusion** technology and user interaction technology are the basic support technology to achieve an AR system.

Due to the perfect combination of virtual and real objects, it is

---

① augmented reality: AR,增强现实。一种将虚拟信息与真实世界巧妙融合的技术,广泛运用了多媒体、三维建模、实时跟踪及注册、智能交互、传感等多种技术手段,将计算机生成的文字、图像、三维模型、音乐、视频等虚拟信息模拟仿真后应用到真实世界中,两种信息互为补充,实现对真实世界的"增强"。

Figure 8B-2　A Example of AR

necessary to combine the virtual objects into the real world. This process is called **registration**.

Some **high-tech** companies are by constructing augmented reality platform, to achieve more extensive application form, such as Wikitude① and Layar②, they have an open SDK, Android, blackberry, let iOS 10 and Web(HTML5, JavaScript and developer support CSS) augmented reality module can create **cross platform** based on geographic location, image recognition, 3D **animation**.

But common standards, interfaces, and communication protocols between and among virtual environment systems are still in development. Several collaborations and working groups have been established in an attempt to create the types of standards and protocols that would be needed to support interoperability between virtual environments.

Ⅲ. What Companies Are Involved?

- **Facebook**

One of the biggest fans of a proposed metaverse has been Facebook③ chief Mark Zuckerberg. He has spoken over the years about how such a world would fit in with his company, a massive

registration
/ˌredʒɪˈstreɪʃn/
n. 注册
high-tech
高科技

cross platform
跨平台
animation
/ˌænɪˈmeɪʃn/
n. 动画

---

① Wikitude：一家专门从事智能手机增强现实应用的厂商，分别出品了两款应用，世界上第一个增强现实导航软件 Wikitude drive 和增强现实浏览器 Wikitude World Browser。
② Layar：全球第一款增强现实感的手机浏览器，由荷兰软件公司 SPRXmobile 研发设计。它能向人们展示周边环境的真实图像，只要将手机的摄像头对准建筑物，就能在手机的屏幕下方看到与这栋建筑物相关的、精确的现实数据。
③ Facebook：脸书，美国的一个社交网络服务网站，总部位于美国加利福尼亚州，创立于 2004 年，创始人为马克·扎克伯格，2021 年 10 月更名为 Meta。

social media service with international reach.

Facebook has backed the idea by investing a lot of money in VR and AR technologies, including the development of **headsets** that promise to create the most realistic virtual interactions possible. The company also announced in July 2021 that it had created a new team to specifically develop metaverse products.

Earlier this year, Zuckerberg called the metaverse the "next generation of the internet and next chapter for us as a company." He said the plans would create "entirely new experiences and economic opportunities".

- Google

Google has also been developing VR and AR tools that aim to "bridge the digital and physical worlds". One of its latest tools is called Google **Lens**. It enables users to use a device's camera to capture an object. The technology then uses image recognition and Google's search system to describe what the object is and provide information about it. Such a system could one day be used with headsets in a metaverse.

- Apple

Apple has also reportedly been working on the development of AR smart glasses. ARKit (see Figure 8B-3), an AR development platform, was launched by Apple in 2017.

Figure 8B-3　ARKit by Apple

- Microsoft

In May 2021, Microsoft explained how it was developing a series of "metaverse Apps" designed to help business users of its

Azure cloud computing service combine virtual and physical elements, see Figure 8B-4.

Figure 8B-4　Azure by Microsoft

Ⅳ. Future Prospects

To **Silicon Valley** dreamers, this immersive, networked, three-dimensional world will eventually succeed the two-dimensional internet that exists today. It requires infrastructure and processing power that does not yet exist. But if you consider Fortnite's[①] expansive world, Facebook's Horizon[②] and other investments by big tech firms in similar products, the metaverse is closer to reality than it was when Mr Stephenson first dreamed it up.

With the development of Web 3.0, 3D Internet[③] emerges. The 3D era may be very similar to the current development status of AR/VR. Take VR as an example. At present, the mainstream VR applications are mainly games and entertainment, with a relatively large number. At the same time, VR in the vertical industry for education, training, medical, cultural and tourism and other fields of exploration and cases, which jointly promote the development

Silicon Valley
硅谷

---

① Fortnite：《堡垒之夜》，大逃杀类现象级热门游戏代表作。2020 年 9 月，NVIDIA 与 Epic Games 携手宣布该游戏支持实时光线追踪、AI 驱动的 DLSS（深度学习超级采样）等突破性技术，使游戏更具美感，互动性更强。
② Facebook's Horizon：Facebook 开源的一个端到端的应用强化学习平台 Horizon，它使用增强学习（Reinforcement Learning，RL）来优化数十亿人使用的产品和服务，旨在促使强化学习从研究论文过渡到生产环境。
③ 3D Internet：三维互联网，新一代互联网技术。在现有的互联网架构下，互联网网站和 App 全面兼容三维技术，三维用户可通过其虚拟人物在三维网页中开展人与人、人与物的深层交互活动。

and landing of VR technology.

Different application scenarios will create different "meta-universes", which will be divided into different domains in the future.

When we talk about the metasverse, we need to see if it has an infrastructure. Only by strengthening social platform, developing VR, AR, **gesture** operation, Brain-computer interface①, EMG② control and other technologies, and continuous technical reserve, the metaverser will get closer and closer.

gesture
/'dʒestʃə/
n. 手势

Exercises

Ⅰ. Fill in the blanks with the information given in the text.

1. The word metaverse combines the prefix "meta" (meaning "_____") with "universe".

2. The Metaverse is the concept of a future iteration of the internet, made up of persistent, shared, decentralized, _____ spaces linked into a perceived virtual universe.

3. The use of computers or other intelligent computing devices to simulate a virtual world of three degrees, providing users with visual, auditory, tactile and other _____ simulation, so that the user as immersive.

4. The use of the most advanced 3D sound solutions to simulate surround auditory experience, allowing users to feel _____.

5. AR is a new technology developed on the basis of virtual reality, also known as _____ reality.

Ⅱ. Translate the following terms or phrases from English into Chinese.

software usability	software design
software agent	three-dimensional space
virtual reality	VR
AR	cloud rendering
audio-visual experience	the plane display mode
a panoramic display	handle manipulation
speech recognition	real scene
positioning technology	

---

① Brain-computer interface：BCI,脑机接口,指在人或动物大脑与外部设备之间创建的直接连接,实现脑与设备的信息交换。

② EMG：肌电信号,是众多肌纤维中运动单元动作电位在时间和空间上的叠加。

Section C Robotics

Ⅰ. Introduction

Although the field of artificial intelligence is relatively young, it has produced some **astonishing** results such as expert chess players, computers that appear to learn and reason, and machines that coordinate their activities to achieve a common goal such as winning a soccer game. In artificial intelligence, today's science fiction might well be tomorrow's reality.

A robot is a mechanical or virtual, artificial **agent**. It is usually a system, which, by its appearance or movements, conveys a sense that it has intent or agency of its own.

While there is still discussion about which machines qualify as robots, a typical robot will have several, though not necessarily all of the following properties.

- It is not "natural", i.e. artificially created.
- It can sense its environment, and manipulate or interact with things in it.
- It has some ability to make choices based on the environment, often using automatic control or a preprogrammed sequence.
- It is programmable.
- It moves with one or more axes of rotation or translation.
- It makes **dexterous** coordinated movements.
- It appears to have intent or agency.

Ⅱ. The History of Robotics

Al-Jazari①, an Arab **Muslim** inventor, designed and constructed a number of automatic machines, including kitchen appliances, musical automata powered by water, and the first programmable **humanoid** robot in 1206.

One of the first recorded designs of a humanoid robot was made by Leonardo da Vinci② in around 1495. Da Vinci's notebooks, rediscovered in the 1950s, contain detailed drawings of a mechanical knight able to sit up, wave its arms and move its head and jaw.

An early **automaton** was created in 1738 by Jacques de Vaucanson③, who created a mechanical duck that was able to eat and digest grain, flap its wings, and excrete.

The first electronic autonomous robots were created by William Grey Walter④ of the Burden Neurological Institute at Bristol, England⑤ in 1948 and 1949. They were named Elmer and Elsie. These robots could sense light and contact with external objects, and use these stimuli to navigate.

The first truly modern robot, digitally operated, programmable, and teachable, was invented by George Devol⑥ in 1954 and was ultimately called the **Unimate**.

Ⅲ. The Development of Robotics

Recently, the development of faster, lighter weight computers has lead to greater research in mobile robots that can move about.

---

① Al-Jazari：加扎里(1136—1206)，阿拉伯人，发明了最重要的机械——抽水机，被称为现代工程之父。
② Leonardo da Vinci：列昂纳多·达·芬奇(1452—1519)，意大利人，欧洲文艺复兴时期的艺术家、科学家、发明家。
③ Jacques de Vaucanson：雅卡尔·德·沃康桑(1709—1782)，德国发明家、艺术家。
④ William Grey Walter：威廉·格雷·沃尔特(1910—1977)，英国人，著名神经系统科学家和机器人学家，在20世纪40年代制造的机器乌龟是世界上最早的具有自主意识的电动机器人。
⑤ the Burden Neurological Institute at Bristol, England：英国布里斯托尔大学的伯登神经学研究所。
⑥ George Devol：乔治·德沃尔(1912—2011)，美国人，发明家，机器人的发明者之一。

Researchers in robot **locomotion** have developed robots that swim like fish, fly like **dragonflies**, hop like **grasshoppers**, and crawl like snakes.

Wheeled robots are limited in the type of **terrain** they can traverse. Using combinations of wheels or tracks to climb stairs or roll over rocks, is the goal of current research. As an example, the **NASA**① **Mars rovers**② (see Figure 8C-1) used specially designed wheels to move on rocky soil.

Figure 8C-1　Mars Rover

Legged robots offer greater mobility but are significantly more complex. For instance, two-legged robots, designed to walk as humans, must constantly monitor and adjust their stance or they will fall. As exemplified by the two-legged humanoid robot named **Asimo**③ (see Figure 8C-2), developed by **Honda**④, that can walk up stairs and even run.

Figure 8C-2　Asimo

---

① NASA: National Aeronautics and Space Administration,美国航空航天局。
② Mars rovers:火星探测漫游者。
③ Asimo:日本本田工业技研公司研制的仿人机器人(阿西莫)。
④ Honda:本田(汽车品牌)。

locomotion
/ˌləʊkəˈməʊʃn/
n. 运动
dragonfly
/ˈdrægənflaɪ/
n. 蜻蜓
grasshopper
/ˈɡrɑːʃɒpə(r)/
n. 蚱蜢,蝗虫
terrain
/ˈtereɪn/
n. 地形

Despite great advances in manipulators and locomotion, most robots are still not very autonomous. Other mobile robots such as the NASA Mars rovers and military **unmanned aerial vehicles** (UAVs) (see Figure 8C-3) rely on human operators for their intelligence.

UAV
无人战斗机

Figure 8C-3　A Unmanned Aerial Vehicle

Overcoming this dependency on humans is a major goal of current research. One question deals with what an autonomous robot needs to know about its environment and to what degree it needs to plan its actions in advance. Research in this direction depends heavily on progress in **knowledge representation** and storage as well as improved **reasoning** and plan development techniques.

The iRobot Roomba① vacuum cleaner (see Figure 8C-4), which moves about a floor in a reactive mode without bothering to remember the details of **furniture** and other obstacles. After all, the family pet will probably not be in the same place next time.

knowledge representation
知识表达
reasoning
/ˈriːznɪŋ/
n. 推理
furniture
/ˈfɜːnɪtʃə(r)/
n. 家具

Figure 8C-4　Roomba

---

① Roomba：美国的家用机器人供应商 iRobot 公司研发的最新一代的智能机器人，是一款机器人吸尘器。

Of course, no single approach will likely prove the best for all situations. Truly autonomous robots will most likely use multiple levels of reasoning and planning, applying high-level techniques to set and achieve major goals and lower-level reactive systems to achieve minor sub-goals. An example of such multilevel reasoning is found in the Robocup① competition—an international competition of robot soccer teams—that serves as a forum for research toward developing a team of robots that can beat world-class human soccer teams by the year 2050.

Another example of research in robotics is the field known as evolutionary robotics in which theories of evolution are applied to develop schemes for both low-level reactive rules and high-level reasoning. The survival-of-the-fittest theory② being used to develop devices that over multiple generations acquire their own means of balance or mobility. For example, the control system for a swimming **tadpole** robot was transferred to a similar robot with legs. Then evolutionary techniques were applied within the control system to obtain a robot that crawled. In other instances, evolutionary techniques have been applied to a robot's physical body to discover positions for sensors that are optimal for performing a particular task. More challenging research seeks ways to evolve software control systems simultaneously with physical body structures.

Ⅳ. Prospect

To list all the impressive results from research in robotics would be an overwhelming task. Our current robots are far from the powerful robots in fictional movies and novels, but they have achieved impressive successes on specific tasks. We have robots that can drive in traffic, behave like pet dogs, and guide **weapons** to their targets. However, while **relishing** in these successes, we should note that the affection we feel for an artificial pet dog and the **awesome** power of smart weapons raise social and ethical

---

① Robocup: 机器人世界杯足球锦标赛,以 MAS(Multi-Agent System)和 DAI (Distributed Artificial Intelligence)为背景,促进人工智能和智能机器人的研究。
② survival-of-the-fittest theory: 适者生存理论。

questions that challenge society. Our future is what we make it.

Exercises

Ⅰ. Fill in the blanks with the information given in the text.

1. In _____, today's science fiction might well be tomorrow's reality.

2. A robot is a mechanical or virtual, artificial _____.

3. Recently, the development of faster, lighter weight computers has lead to greater research in _____ robots that can move about.

4. Despite great advances in manipulators and locomotion, most robots are still not very _____.

5. Another example of research in robotics is the field known as _____, where the survival-of-the-fittest theory being used to develop devices that over multiple generations acquire their own means of balance or mobility.

6. More challenging research seeks ways to evolve software control systems simultaneously with physical _____ structures.

Ⅱ. Translate the following terms or phrases from English into Chinese.

artificial intelligence   two-legged robot
humanoid robot   wheeled robot
unmanned aerial vehicle   knowledge representation

Unit Nine Computer Security

Section A Network Security

When a computer is connected to a network, it becomes subject to unauthorized access and **vandalism**. There are some topics associated with these problems.

Ⅰ. Forms of Attack

There are numerous ways that a computer system and its contents can be attacked via network connections. Many of these incorporate the use of malicious software. They are specifically designed to damage or disrupt a computer system. Such as viruses, worms, Trojan horses, spyware and so on.

A computer virus is a program or piece of code that is loaded onto your computer without your knowledge and runs against your wishes. Viruses can also replicate themselves. All computer viruses are man-made. So—called benign viruses might simply display a message. Malignant viruses are designed to damage the system. The attack is to wipe out data, to delete files, or to format the hard disk.

A computer worm is a standalone **malware** computer program that **replicates** itself in order to spread to other computers. Often, it uses a computer network to spread itself. Unlike a computer virus, it does not need to attach itself to an existing program. It fills a

vandalism
/ˈvændəlɪzəm/
n. 故意破坏；捣毁

malware
恶意软件；流氓软件
replicate
/ˈreplɪkeɪt/
vt. 复制

computer system with self-replicating information, clogging the system so that its operations are slowed or stopped.

A Trojan horse is a program in which **malicious** or harmful code is contained inside apparently harmless programming or data in such a way that it can get control and do its chosen form of damage, such as ruining the **file allocation table** on your hard disk.

Spyware is any technology that aids in gathering information about a person or organization without their knowledge, it is sometimes called a spybot or tracking software. Spyware covertly gathers user information through the user's Internet connection without his or her knowledge, usually for advertising purposes. Spyware applications are typically **bundled** as a hidden component of freeware or shareware programs that can be downloaded from the Internet. Once installed, the spyware monitors user activity on the Internet and transmits that information in the background to someone else. Spyware can also gather information about E-mail addresses and even passwords and credit card numbers.

In contrast to suffering from such internal infections as viruses and spyware, a computer in a network can also be attacked by software being executed on other computers in the system. An example is a **denial of service(DoS)** attack, which is the process of **overloading** a computer with messages. Many DoS attacks, such as the Ping of Death and Tear Drop attacks, exploit limitations in the TCP/IP protocols. DoS attacks have been launched against large commercial Web servers on the Internet to disrupt the company's business and in some cases have brought the company's commercial activity to a halt.

Another problem associated with an abundance of unwanted messages is **spam**. Spam is technically a way of sending out bulk messages via electronic means and without permission from the receiver. It is in the category of junk mail.

Ⅱ. Protections

- Firewall

Firewall technology emerged in the late 1980s when the Internet was a fairly new technology in terms of its global use and

---

malicious
/məˈlɪʃəs/
adj. 恶意的

file allocation table
文件分配表

bundle
/ˈbʌndl/
vt. 捆绑

denial of service
拒绝服务
overload
/ˌəʊvəˈləʊd/
vt. 使超负荷

spam
/spæm/
n. 垃圾邮件

firewall
/ˈfaɪəwɔːl/
n. 防火墙

connectivity. In order to provide some level of separation between an organization's intranet and the Internet, firewalls have been employed. A firewall is simply a group of components that collectively from a barrier between two networks. A firewall might be installed at the gateway of an organization's intranet to filter messages passing in and out of the region. Such firewalls might be designed to block outgoing messages with certain destination addresses or to block incoming messages from origins that are known to be sources of trouble. Conceptually, there are three types of firewalls: network layer, application layer, and hybrids. These days, most firewalls fall into the "hybrid" category, which do network filtering as well as some amount of application inspection. The amount changes depending on the vendor, product, protocol and version, so some level of digging and/or testing is often necessary.

- **Auditing Software**

Using network **auditing** software, a system administrator can detect a sudden increase in message traffic at various locations within the administrator's **realm**, monitor the activities of the system's firewalls, and analyze the pattern of requests being made by the individual computers in order to detect irregularities. In effect, auditing software is an administrator's primary tool for identifying problems before they grow out of control.

audit
/ˈɔːdɪt/
vt. 审计;审核
realm
/relm/
n. 领域;范围

- **Antivirus Software**

Antivirus software is an extremely important tool to help detect and block attempts by the "bad guys" to infect our computer. A new virus must first infect some computers before it is discovered. Thus, antivirus software must be routinely maintained by downloading updates from the software's **vendor**. Even this, however, does not guarantee the safety of a computer. Thus, a wise computer user never opens email attachments from unfamiliar sources, does not download software without first confirming its reliability, does not respond to pop-up adds, and does not leave a PC connected to the Internet when such connection is not necessary. There are new viruses and attack methods being generated by the hour today. The virus definitions of the system

antivirus software
防病毒软件

vendor
/ˈvendə(r)/
n. 供应商;卖主

have to be updated very regularly for protecting your system form the latest viruses. This updating should be done automatically as it can be forgotten all too easily if it is manual.

- **Encryption**

Another means of protecting information is **encryption**. It refers to algorithmic schemes that encode plain text into non-readable form or **ciphertext**, providing privacy. The receiver of the encrypted text uses a "key" to decrypt the message, returning it to its original plain text form. The key is the trigger mechanism to the algorithm.

Today, many traditional Internet applications have been altered to incorporate encryption techniques. Examples include **FTPS**[①], which is a secure version of FTP, and **SSH**[②], and the secure version of HTTP, known as **HTTPS**[③], which is used by most financial institutions to provide customers with secure Internet access to their accounts. The backbone of HTTPS is the protocol system known as Secure Sockets Layer(SSL[④]).

There are two main types of encryption: **asymmetric** encryption (also called **public-key encryption**) and symmetric encryption. The former uses one key to encrypt a message and another to decrypt the message, while the latter uses the same key to encrypt and decrypt the message.

There are many types of encryption and not all of them are reliable. The same computer power that yields strong encryption can be used to break weak schemes. Initially, 64b encryption was thought to be quite strong, but today 128b, even 256b is the standard. To be secure, the recipient of the data—often a server—must be positively identified as being the approved party. This is usually accomplished online using **digital signatures** or **certificates**.

As more people realize the open nature of the Internet, email and instant messaging, encryption will undoubtedly become more popular. Without it, information passed on the Internet is not only

---

① FTPS：一种多传输协议，相当于加密版的 FTP，也称作 FTP-SSL 和 FTP-over-SSL。
② SSH：secure shell，安全壳协议，为一项创建在应用层和传输层基础上的安全协议。
③ HTTPS：以安全为目标的 HTTP 通道，简单讲是 HTTP 的安全版，即 HTTP 下加入 SSL 层，HTTPS 的安全基础是 SSL，因此加密的详细内容就需要 SSL。
④ SSL：安全套接层协议，是网景公司于 1994 年提出的互联网信息加密传输协议。

encryption
/enˈkrɪpʃən/
n. 编密码；加密
ciphertext
/ˈsaɪfətekst/
n. 密文

asymmetric
/ˌeɪsɪˈmetrɪk/
adj. 不对称的
public-key encryption
公钥加密

digital signature
数字签名
certificate
/səˈtɪfɪkət/
n. 证书

available for virtually anyone to snag and read, but is often stored for years on servers that can change hands or become compromised in any number of ways. For all of these reasons, it is a goal worth pursuing.

Exercises

I. Fill in the blanks with the information given in the text.

1. When a computer is connected to a network, it becomes subject to _____ access and _____.

2. Viruses can also _____ themselves and are man-made.

3. A computer worm is a standalone _____ computer program that replicates itself in order to spread to other computers.

4. A Trojan horse is contained inside apparently _____ programming or data.

5. _____ can also gather information about E-mail addresses and even passwords and credit card numbers.

6. _____ is technically a way of sending out bulk messages via electronic means and without _____ from the receiver.

7. A(an) _____ is simply a group of components that collectively from a barrier between two _____.

8. _____ is an extremely important tool to help detect and block attempts by the "bad guys" to infect our computer.

9. There are two main types of encryption: _____ and symmetric encryption.

10. Today, many traditional Internet applications have been altered to incorporate encryption techniques. Examples include _____ and _____.

II. Translate the following terms or phrases from English into Chinese.

malicious software	Trojan horse
spyware	file allocation table
DoS	spam
firewall	asymmetric
vendor	antivirus software
encryption	ciphertext
public-key encryption	private key
certificate	digital signature
gateway	virus
worm	password

III. Translate the following passage from English into Chinese.

Computer Virus

Computer viruses are just one example of what is commonly referred to as malicious code or malicious programs. Malicious programs are created to perform a series of harmful actions on a computer system. Examples of some actions include file deletion, file corruption, data theft, and the less harmful but equally annoying practical joke. These programs often remain dormant and hidden until an activation event occurs. Examples of activation events are program execution and specific access dates such as March 15, system reboot, and file access. When the predetermined activation event occurs, the malicious program begins its task.

Section B Digital Signature

Ⅰ. Introduction

A digital signature is a mathematical scheme for demonstrating the **authenticity** of a digital message or document. A valid digital signature gives a recipient reason to believe that the message was created by a known sender, the sender cannot deny having sent the message and that the message was not altered in transit. Digital signatures rely on certain types of encryption to ensure authentication. Encryption is the process of taking all the data that one computer is sending to another and **encoding** it into a form that only the other computer will be able to **decode**. Authentication is the process of verifying that information is coming from a trusted source.

Ⅱ. Digital Signatures and Ink Signatures

Digital signatures are equivalent to traditional ink signatures in many respects. An ink signature could be replicated from one document to another by copying the image manually or digitally, but to have credible signature copies that can resist some scrutiny is a significant manual or technical skill, and to produce ink signature copies that resist professional scrutiny is very difficult. Digital signatures cryptographically bind an electronic identity to an electronic document and the digital signature cannot be copied to another document. Paper contracts sometimes have the ink signature block on the last page, and the previous pages may be replaced after a signature is applied. Digital signatures can be applied to an entire document, such that the digital signature on the last page will indicate tampering if any data on any of the pages have been altered, but this can also be achieved by signing with ink and numbering all pages of the contract.

Digital signatures are commonly used for software distribution, financial transactions, and in other cases where it is important to detect forgery or tampering.

Ⅲ. Asymmetric Cryptography

Digital signatures employ a type of asymmetric cryptography. For messages sent through a nonsecure channel, a properly implemented digital signature gives the receiver reason to believe the message was sent by the claimed sender.

Asymmetric cryptography, also called Public-key cryptography, is a class of cryptographic algorithms which requires two separate keys, one of which is secret(or private) and one of which is public. Although different, the two parts of this key pair are mathematically linked. The public key is used to encrypt **plaintext** or to verify a digital signature; whereas the private key is used to decrypt ciphertext or to create a digital signature. The term "asymmetric" stems from the use of different keys to perform these opposite functions, each the inverse of the other—as contrasted with conventional("symmetric") cryptography which relies on the same key to perform both.

plaintext
/pleɪnˈtekst/
n. 明码文本

In Digital signatures a message is signed with the sender's private key and can be verified by anyone who has access to the sender's public key. This **verification** proves that the sender had access to the private key, and therefore is likely to be the person associated with the public key. This also ensures that the message has not been tampered, as any manipulation of the message will result in changes to the encoded message digest, which otherwise remains unchanged between the sender and receiver.

verification
/ˌverɪfɪˈkeɪʃn/
n. 证明；证实

Ⅳ. Digital Certificates

To implement public key encryption on a large scale, such as a secure Web server might need, requires a different approach. This is where digital certificates come in. A digital certificate is essentially a bit of information that says the Web server is trusted by an independent source known as a **Certificate Authority**. The Certificate Authority acts as the middleman that both computers trust. It confirms that each computer is in fact who they say they are and then provides the public keys of each computer to the other.

Digital certificates have two basic functions. The first is to certify that the people, the website, and the network resources such as servers and routers are reliable sources, in other words, who or what they claim to be. The second function is to provide protection for the data exchanged from the visitor and the website from tampering or even theft, such as credit card information.

A digital certificate contains the name of the organization or individual, the business address, digital signature, public key, serial number, and expiration date. When you are online and your web browser attempts to secure a connection, the digital certificate issued for that website is checked by the web **browser** to be sure that all is well and that you can browse securely. The web browser basically has a built in list of all the main certification authorities and their public keys and uses that information to decrypt the digital signature. This allows the browser to quickly check for problems, abnormalities, and if everything checks out the secure connection is enabled. When the browser finds an expired certificate or mismatched information, a dialog box will pop up with an alert.

Certificate Authority
证书授权中心

browser
/ˈbraʊzə(r)/
n. 浏览器

Exercises

Ⅰ. Fill in the blanks with the information given in the text.

1. A digital signature is a mathematical scheme for demonstrating the _____ of a digital message or document.

2. Digital signatures rely on certain types of _____ to ensure authentication.

3. Encryption is the process of taking all the data that one computer is sending to another and _____ it into a form that only the other computer will be able to _____.

4. Digital signatures are commonly used for software _____, financial transactions, and in other cases where it is important to detect _____.

5. The public key is used to encrypt plaintext or to _____ a digital signature;

whereas the private key is used to decrypt ciphertext or to _____ a digital signature.

6. A digital certificate is essentially a bit of information that says the Web server is trusted by an independent source known as a _____.

7. A digital certificate contains the name of the organization or individual, the business address, _____, public key, serial number, and expiration date.

Ⅱ. Translate the following terms or phrases from English into Chinese.

Digital Signature	plaintext
Certificate Authority	router
browser	middleman
dialog box	

Section C Smartphone Security

I. How Safe is Your Smartphone

Smartphones are getting pretty clever these days, but it is unlikely they will **outwit** the cybercriminals as **fraudsters** increasingly go mobile.

Android Market, the shop front for applications aimed at Android smartphones, once was hit by around 60 malicious apps. It is thought that they did little real damage other than to Android's reputation, but the incident put the issue of mobile security back in the headlines. The truth is you're packing a lot of sensitive information on your phone, and you should keep it safe.

Phones are attractive to criminals because they are essentially mini computers but with some important added extras. Phones also have direct access to address books, calendars as well as offering an ability to generate revenue. The type of personal data typically stored on a phone opens up a rich new vein for the modern fraudster's preferred crime—identity theft. However, a more immediate income can be made from so-called **rogue** dialling programs—malicious bits of code capable of placing calls, unbeknown to the owner.

They are, according to Ovum analyst Graham Titterington, the "number one malware threat" to smartphones. "Rogue dialling connects the phone automatically to a premium number that invariably belongs to a crook based in another country," he explained. But it is not an **insurmountable** issue, he thinks. "I don't understand why the mobile operators can't just cut off payments—

outwit
/ˌaʊtˈwɪt/
vt. 以智取胜；以计击败

fraudster
/ˈfrɔːdstə(r)/
n. 行骗者

rogue
/rəʊɡ/
n. 流氓；无赖

insurmountable
/ˌɪnsəˈmaʊntəbl/
adj. 不可逾越的

then the problem goes away. But this type of international co-operation seems to be lacking at the moment", said Mr. Titterington. The close relationship between smartphones and location poses a risk that malicious apps will be able to track exactly where a person is at any given time.

Android may have hit the headlines, but all smartphone operating systems have been targeted by malware of one kind or another.

To date, most iPhone security lapses have focused on offering users the power to break free from Apple's control with software that **"jailbreaks"** the iPhone, a modification which enables users to run non-Apple approved software.

Ⅱ. Mobile Malwares

- Android-DroidDream: the most recent and most advanced piece of malware hit apps and allowed product ID and user ID of phone to be transmitted to remote server.
- Zeus-in-the-mobile: a trojan working with the Windows virus Zeus, affecting Symbian and Blackberry handsets and aiming to steal online banking details.
- Android-Geinimi: similar to the market app attack, it took official apps, added malware and released them via Asian app markets. Could send SMSs, harvest phone data and make phone calls.
- Android-ADRD: another trojan that pirated official Android Apps.

Blackberry handsets and Symbian phones have been targeted by a mobile version of the Zeus trojan. **Victims** were directed to a fake website where they are invited to download an app which then steals their banking details. Such **phishing** attacks are likely to become a huge problem for smartphones, because you can't always see the whole screen and you might be more likely to click on things you wouldn't click on a computer screen. And when mobile banking reaches a critical mass, there will be a good reason for criminals to phish from mobiles. There needs to be a financial **incentive** and that incentive isn't there right now, but consumers definitely want more service on their mobiles, like electronic

jailbreak
/'dʒeɪlbreɪk/
n. 越狱
vt. (对手机,特别是iPhone 等电子产品所做的)破解

victim
/'vɪktɪm/
n. 受害者;受骗者
phishing
/'fɪʃɪŋ/
n. 网络仿冒;网络钓鱼

incentive
/ɪn'sentɪv/
n. 刺激;鼓励

wallets and banking, so the potential is huge.

There are various ways to attack a mobile phone but by far the most popular is through downloadable applications. Some experts think that Android's Marketplace is especially **vulnerable** because it is more open than Apple and Microsoft's systems.

III. Safety Steps

It was hard for application stores to separate programs using personal information legitimately from those with a malicious intent. Many handset hackers would likely copy existing applications and add-in malicious code. It's way less effort to hack into someone else's application, as you do not have to write it yourself. Many would do that, to ensure they hit plenty of victims. What's most important for hackers is how do they get scale. If they write their own application, such as a game, they may only get 200 downloads. By contrast, stealing a popular application, packing it with **booby-trapped code** and offering it for free can reap rewards. Some application makers have found that 97% of the people using their software are doing so via **pirated versions**.

Application stores are making efforts to police the programs they offer. So far the number of booby-trapped applications remains low. But many feel the threat is only likely to grow. Users can take a few simple steps to stay safe.

- Phone owners should also back up data on their handsets to a PC or net-based service to guard against problems.
- Always lock your device, don't leave it lying around open. That way a password will need to be entered to activate the device.
- A strong password includes a combination of upper and lower caseletters and numbers. Never store passwords on your phone and keep your password a secret!
- Keep the software on your phone up to date. Most manufacturers offer updates regularly which often include security updates.
- Try not to use unsecured WiFi networks in public areas to access your email, purchase things or do your banking online. If a network is easy to access by you, then it will be

vulnerable
/ˈvʌlnərəbl/
adj. 易受攻击的；
易受伤的

booby-trapped code
诡雷代码

pirated version
盗版

easy to access by potential hackers.
- Take care when downloading apps—if something looks too good to be true, it probably is! Likewise, don't click on links unless from a trusted source, and be particularly careful if you unexpectedly receive a link to a banking website.

Exercises

Ⅰ. Fill in the blanks with the information given in the text.

1. Smartphones are getting pretty clever these days but it is unlikely they will _____ the cybercriminals as _____ increasingly go mobile.

2. Android Market, the shop front for applications aimed at Android _____, once was hit by around 60 _____ Apps.

3. _____ connects the phone automatically to a premium number that invariably belongs to a crook based in another country.

4. Android may have hit the _____ but all smartphone operating systems have been targeted by malware of one kind or another.

5. To date, most iPhone security lapses have focused on offering users the power to break free from Apple's control with software that "_____" the iPhone, a modification which enables users to run non-Apple approved software.

6. There are various ways to _____ a mobile phone but by far the most popular is through downloadable _____.

7. A strong password includes a combination of _____ case letters and numbers.

Ⅱ. Translate the following terms or phrases from English into Chinese.

smartphone   android market
rogue dialling program   malware
jailbreak   phishing
booby-trapped code   pirated version
victim   online bank

词 汇 表

A

academic /ˌækəˈdemɪk/ adj. 学术的
accelerometer /ækˌseləˈrɒmɪtə/ n. 加速器
acceptability /əkˌseptəˈbɪlətɪ/ n. 可接受性
acquisition /ˌækwɪˈzɪʃn/ n. 获得；取得
actuator /ˈæktʃʊeɪtə/ n. 执行器
addressable /əˈdresəbl/ adj. 可寻址的
advocacy /ˈædvəkəsɪ/ n. 支持；辩护
affordability /əˌfɔːdəˈbɪlətɪ/ n. 可购性；成本合理性
agent /ˈeɪdʒənt/ n. 代理
aggregate /ˈægrɪgət/ v. 聚合；聚集
agile method 敏捷方法
agile /ˈædʒaɪl/ adj. 灵活的；敏捷的
alert /əˈlɜːt/ n. 警示信息
AlphaGo 阿尔法围棋
algorithmic /ˌælgəˈrɪðmɪk/ adj. 算法的
allocate /ˈæləkeɪt/ vt. 分配
allocation /ˌæləˈkeɪʃn/ n. 分配
alternative /ɔːlˈtɜːnətɪv/ n. 二中择一；可供选择的事物
ambient /ˈæmbɪənt/ adj. 周围的
ambiguous /æmˈbɪgjuəs/ adj. 有歧义的
amplification /ˌæmplɪfɪˈkeɪʃn/ n. 放大
analytical /ˌænəˈlɪtɪkəl/ adj. 分析的
animation /ˌænɪˈmeɪʃn/ n. 动画
anonymous /əˈnɒnɪməs/ adj. 匿名的
antenna /ænˈtenə/ n. 天线
antivirus software 防病毒软件

API 应用程序编程接口
applet /ˈæplɪt/ n. Java的小应用程序
appliance /əˈplaɪəns/ n. 家用电器
archival /ɑːˈkaɪvəl/ adj. 档案的
arithmetic /əˈrɪθmətɪk/ n. 算术；计算
arrow key 箭头键
artifact /ˈɑːtɪfækt/ n. 人工制品
artificial /ˌɑːtɪˈfɪʃl/ adj. 人工的
assembly language 汇编语言
asset /ˈæset/ n. 资产
association analysis 关联分析
astonishing /əˈstɒnɪʃɪŋ/ adj. 惊人的
asymmetric /ˌeɪsɪˈmetrɪk/ adj. 不对称的
a trusted third party 可信第三方
audio-visual 视听
auditing /ˈɔːdɪtɪŋ/ vt. 审计；审核
augment /ɔːgˈment/ vt. 增强
authentication /ɔːˌθentɪˈkeɪʃn/ n. 认证；鉴别
authenticity /ˌɔːθenˈtɪsətɪ/ n. 可靠性；真实性
automaton /ɔːˈtɒmətən/ n. 自动机；机器人
Autobahn /ˈɔːtəʊbɑːn/ n. 德国高速公路
autonomous /ɔːˈtɒnəməs/ adj. 自主的
avatar /ˈævətɑː/ n. 虚拟人物
awesome /ˈɔːsəm/ adj. 可怕的；令人惊叹的

B

backbone /ˈbæckəʊn/ n. 骨干
back-end 后端
bar code 条形码
barrier /ˈbæriə/ n. 障碍
Bayesian network 贝叶斯网络
Beyond /bɪˈjɒnd/ prep. 超越
binary /ˈbaɪnəri/ n. 二进制
Bioinformatics /ˌbaɪəʊɪnfəˈmætɪks/ n. 生物信息学
Biometrics /ˌbaɪəʊˈmetrɪks/ n. 生物统计学
blade /bleɪd/ n. 刀片（服务器）
blog /blɒg/ n. 博客
bluetooth /ˈbluːtuːθ/ n. 蓝牙
Blu-ray 蓝光
booby-trapped code 诡雷代码
boolean /ˈbuːliən/ adj. 布尔的
breadth /bredθ/ n. 宽度；广泛
bridge /brɪdʒ/ n. 网桥
browser /ˈbraʊzə(r)/ n. 浏览器
bug /bʌg/ n. (软件)错误；漏洞
bug-free 没有错误的
builder /ˈbɪldə(r)/ n. 生成器
built-in 嵌入的；内置
bundle /ˈbʌndl/ vt. 捆绑
bus /bʌs/ n. 总线
buzzword /ˈbʌzwɜːd/ n. 时髦术语
bytecode /bɪtiːˈkəʊd/ n. 字节码

C

cabinet /ˈkæbɪnɪt/ n. 橱柜；箱
cable radio 无线电
calculus /ˈkælkjələs/ n. 微积分；演算
California /ˌkæləˈfɔːniə/ n. 加利福尼亚
capacity /kəˈpæsəti/ n. 容量；能力
categorize /ˈkætɪɡəraɪz/ vt. 分类
centralized /ˈsentrəlaɪzd/ adj. 集中的
Certificate Authority 证书授权中心
certificate /səˈtɪfɪkət/ n. 证书
chamber /ˈtʃeɪmbə/ n. 房间
channel /ˈtʃænl/ n. 通道；频道
chassis /ˈʃæsi/ n. 底盘；底架
chip /tʃɪp/ n. 芯片
chore /tʃɔː/ n. 杂事
ciphertext /ˈsaɪfətekst/ n. 密文
circuit /ˈsɜːkɪt/ n. 电路
claim /kleɪm/ n. 索赔
class /klɑːs/ n. 类
class-based 基于类的
clause /klɔːz/ n. 条款
CLI 命令行界面
client/server 客户/服务器
cloud computing 云计算
cluster analysis 聚类分析
cluster /ˈklʌstə/ n. 集群
coincidence /kəʊˈɪnsɪdəns/ n. 巧合；一致
Colossus /kəˈlɒsəs/ n. 巨人
compass /ˈkʌmpəs/ n. 指南针；罗盘
compatible /kəmˈpætəbl/ adj. 兼容的
compensation /ˌkɒmpenˈseɪʃn/ n. 补偿；赔偿
compiler /kəmˈpaɪlə(r)/ n. 编译器
computing power 计算能力
concurrency /kənˈkʌrənsi/ n. 并发性
confidential /ˌkɒnfɪˈdenʃl/ adj. 机密的

conflict /'kɒnflɪkt/ n. 冲突
conglomerate /kən'glɒmərɪt/ n. 集团；聚块
commercialization /kəˌmɜːrʃələ'zeɪʃn/ n. 商业化；商品化
conditional probability 条件概率
consensus mechanism 共识机制
consideration /kənˌsɪdə'reɪʃn/ n. 考虑；考察
console /kən'səʊl/ n. 控制台
consolidate /kən'sɒlɪdeɪt/ v. 整合
consumption /kən'sʌmpʃn/ n. 消耗；消费
contemporary /kən'temprəri/ adj. 现代的
contention /kən'tenʃn/ n. 竞争
continuum /kən'tɪnjuəm/ n. 连续

contractual /kən'træktʃʊəl/ adj. 合同的
contradiction /ˌkɒntrə'dɪkʃən/ n. 矛盾
coordinate /kəʊ'ɔːdɪnɪt/ v. 使协调
copyright /'kɒpɪraɪt/ n. 版权；著作权
couple /'kʌpl/ v. 连接
coupon /'kuːpɒn/ n. 优惠券；礼券
CPU 中央处理器
crack up 吹捧；赞扬
CRM 客户关系管理
cross platform 跨平台
cross-reference 交叉引用
cryptocurrency /'krɪptəʊkʌrənsi/ n. 数字货币
cryptographic /ˌkrɪptə'græfɪk/ adj. 加密的

D

Daimler /'deɪmlər/ n. 戴姆勒（德国汽车制造商）
data deluge 海量数据
data mining 数据挖掘
data science 数据科学
data warehouse 数据仓库
debugger /ˌdiː'bʌgə/ n. 调试工具
decentralized /ˌdiː'sentrəlaɪzd/ adj. 分散的；去中心化的
decentralized network 去中心网络
deception /dɪ'sepʃn/ n. 欺骗
decision tree 决策树
decode /ˌdiː'kəʊd/ vt. 译码
decouple /ˌdiː'kʌpl/ n. 解耦
dedicated /'dedɪkeɪtɪd/ adj. 专用的
deluge /'deljuːdʒ/ n. 洪水
demographic /ˌdemə'græfɪk/ n. 用户信息统计
demonstration /ˌdemən'streɪʃn/ n. 演示；论证

DBMS 数据库管理系统
denial of service 拒绝服务
deployment /dɪ'plɔɪmənt/ n. 部署；调度
desktop /'desktɒp/ n. 台式机
deterministic /dɪˌtɜːmɪ'nɪstɪk/ adj. 确定性的
deviation /ˌdiːvɪ'eɪʃn/ n. 偏离
devote /dɪ'vəʊt/ vt. 投入于；贡献
dexterous /'dekstrəs/ adj. 灵巧的；熟练的
dictate /dɪk'teɪt/ v. 控制；命令
difference /'dɪfərəns/ n. 差数；差别
digital /'dɪdʒɪtl/ adj. 数字的
digital currency 数字货币
digital signature 数字签名
discipline /'dɪsɪplɪn/ n. 学科；纪律
disk pack 磁盘组
dismissal /dɪs'mɪsl/ n. 解雇；撤退
dispatcher /dɪ'spætʃə/ n. 调度程序
distributed /dɪ'strɪbjuːtɪd/ adj. 分布

式的
do-it-yourself   自助式
downtime   /ˈdaʊntaɪm/   n. 宕机；停工期

dragonfly   /ˈdrægənflaɪ/   n. 蜻蜓
draw   /drɔː/   n. 平局
dynamically   /daɪˈnæmɪkli/   adv. 动态地

E

e-commerce   电子商务
ecosystem   /ˈiːkəʊ sɪstəm/   n. 生态系统
editor   /ˈedɪtə/   n. 编辑程序；编辑器
elasticity   /ˌelæsˈtɪsəti/   n. 弹性部署
electrical heater   电热器
electronic cash   电子现金
electronic payment   电子支付
electronics   /ɪˌlekˈtrɒnɪks/   n. 电子学；电子器件
embed   /ɪmˈbed/   v. 嵌入
embedded system   嵌入式系统
empirical   /ɪmˈpɪrɪkl/   adj. 经验的
encase   /ɪnˈkeɪs/   vt. 包装；装入
encode   /ɪnˈkəʊd/   vt. 译成密码
encompass   /ɪnˈkʌmpəs/   vt. 围绕
encryption   /enˈkrɪpʃn/   n. 编密码；加密
endeavor   /ɪnˈdevə/   n. 努力
ENIAC   埃尼阿克
entanglement   /ɪnˈtæŋglmənt/   n. 纠缠；缠绕物

enthusiast   /ɪnˈθjuːziæst/   n. 热心者；热情者
entity   /ˈentɪti/   n. 实体
escort   /ˈeskɔːt/   v. 保驾护航
Ethernet   /ˈiːθənet/   n. 以太网
ethical   /ˈeθɪkl/   n. 道德的；民族的
Euro   /ˈjʊə rəʊ/   n. 欧元
evaluate   /ɪˈvæljueɪt/   v. 评估
evangelist   /ɪˈvændʒɪlɪst/   n. 传道者
exabyte   /eksəˈbaɪt/   艾字节
exaflop   /ɪɡˈzæflɒps/   n. 百亿亿次
exascale   /ɪɡˈzæskeɪl/   n. 百亿亿次级
execution   /ˌeksɪˈkjuːʃn/   n. 执行
exert   /ɪɡˈzɜːt/   vt. 发挥；行使
explicitly   /ɪkˈsplɪsɪtli/   adv. 明确地
exploratory testing   探索性测试
extract   /ˈekstrækt/   vt. 提取
extraordinary   /ɪkˈstrɔːdnri/   n. 非凡的事
extreme programming   极限编程

F

facet   /ˈfæsɪt/   n. 方面
familiarity   /fəˌmɪliˈærəti/   n. 熟悉；精通
fibre   /ˈfaɪbə/   n. 光纤
fidelity   /fɪˈdeləti/   n. 保真度
fierce   /fɪəs/   adj. 激烈的
figure out   计算出；解决
file allocation table   文件分配表
fine-grained   /faɪnˈɡreɪnd/   adj. 细粒度的

firewall   /ˈfaɪəwɔːl/   n. 防火墙
firmware   /ˈfɜːmweə(r)/   n. 固件
flash drive   闪存驱动器
flat   /flæt/   adj. 单一的
flexible   /ˈfleksɪbl/   adj. 灵活的
flip-flops   n. 正反器
fluctuation   /ˌflʌktjuˈeɪʃn/   n. 波动
formula   /ˈfɔːmjələ/   n. 公式
forward   /ˈfɔːwəd/   vt. 转发

fraudster /ˈfrɔːdstə(r)/ n. 行骗者
free-wheeling /ˌfriːˈwiːlɪŋ/ 单向转动
fungible /ˈfʌndʒəbl/ adj. 可替代

furniture /ˈfɜːnɪtʃə(r)/ n. 家具
fusion /ˈfjuːʒn/ n. 融合
fuzzy logic 模糊逻辑

G

gateway /ˈgeɪtweɪ/ n. 网关
Gb abbr. 千兆位
gear /gɪə/ n. 齿轮
genealogy /ˌdʒiːnɪˈælədʒɪ/ n. 系统学；系谱
genetic /dʒəˈnetɪk/ adj. 遗传的
Genomics /ˌdʒiːˈnəʊmɪks/ n. 基因学

genuinely /ˈdʒenjʊɪnli/ adv. 真正地
gesture /ˈdʒestʃə/ n. 手势
gigabit /ˈgɪgəbɪt/ n. 千兆位
Go /gəʊ/ n. 围棋
grasshopper /ˈgrɑːshɒpə(r)/ n. 蚱蜢,蝗虫
grid /grɪd/ n. 网格

H

handheld /ˌhændˈheld/ adj. 手提式的
handle /ˈhændl/ n. 手柄
headset /ˈhedset/ n. 耳机
Helsinki /ˈhelsɪŋkɪ/ n. 赫尔辛基(芬兰首都)
heterogeneity /ˌhetərəˈdʒɪˈnɪətɪ/ n. 异构性
heuristics /hjuˈrɪstɪks/ n. 启发式
hierarchical /ˌhaɪəˈrɑːkɪkl/ adj. 分层的

high-tech 高科技
host /həʊst/ n. 主机
house /haʊs/ v. 安置
humanoid /ˈhjuːmənɔɪd/ adj. 有人的特点的
hybrid cloud 混合云
hybrid /ˈhaɪbrɪd/ adj. 混合的
hypertext /ˈhaɪpətekst/ n. 超文本

I

IaaS 基础架构即服务
IDE 集成开发环境
identically /aɪˈdentɪkəlɪ/ adv. 同一地；相等地
illusion /ɪˈluːʒn/ n. 错觉；假象
IM 即时消息
imitate /ˈɪmɪteɪt/ v. 模仿
immersive /ɪˈmɜːsɪv/ adj. 沉浸感
immutable /ɪˈmjuːtəbl/ adj. 不可改变的
imperative /ɪmˈperətɪv/ adj. 重要的

inability /ˌɪnəˈbɪlətɪ/ n. 无能为力
inadequate /ɪnˈædɪkwɪt/ adj. 不够好；不足的
incarnation /ˌɪnkɑːˈneɪʃn/ n. 化身
incentive /ɪnˈsentɪv/ n. 刺激；鼓励
incremental model 增量模型
infinitely /ˈɪnfɪnətli/ adv. 无限地
informative /ɪnˈfɔːmətɪv/ adj. 信息量大的
infrared /ˌɪnfrəˈred/ n. 红外线
infrastructure /ˈɪnfrəˌstrʌktʃə/ n. 基础

设施
inherent /ɪnˈhɪərənt/ adj. 固有的；内在的
inherently /ɪnˈherəntli/ adv. 固有地
in-house operating system 自主研发的操作系统
initiate /ɪˈnɪʃieɪt/ v. 发起
inner /ˈɪnə/ adj. 内部的
Inspur /ɪnˈspə/ n. 浪潮集团
instance /ˈɪnstəns/ n. 实例
instant /ˈɪnstənt/ n. 瞬间
instruction /ɪnˈstrʌkʃn/ n. 指令
insurmountable /ˌɪnsəˈmaʊntəbl/ adj. 不可逾越的
integrated /ˈɪntɪgreɪtɪd/ adj. 集成的
integration /ˌɪntɪˈgreɪʃn/ n. 集成；融合
integrator /ˈɪntɪgreɪtə/ n. 积分器
integrity /ɪnˈtegrəti/ n. 完整性
intelligence /ɪnˈtelɪdʒəns/ n. 智力；智慧
intelligent home 智能家居
intelligent terminal 智能终端

intent /ɪnˈtent/ n. 目的
interface /ˈɪntəfeɪs/ n. 接口；界面
intermediate /ˌɪntəˈmiːdiət/ adj. 中间的；中级的
Internet of Things 物联网
interoperability /ˌɪntərˌɒpərəˈbɪləti/ n. 互操作性
interprocess /ˌɪntəˈprɒses/ n. 进程间
interrogation /ɪnˌterəˈgeɪʃn/ n. 询问
intervention /ˌɪntəˈvenʃn/ n. 介入；干预
interwoven /ˌɪntəˈwəʊvən/ v. 交织
invoke /ɪnˈvəʊk/ vt. 调用
IOS 互联网操作系统
IP 网际协议
irregularity /ɪˌregjʊˈlærəti/ n. 不规则
irreversible /ˌɪrɪˈvɜːsəbl/ adj. 不可逆转的
isolated /ˈaɪsəleɪtɪd/ adj. 孤立的
IT 信息技术
iteration /ˌɪtəˈreɪʃn/ n. 迭代次数
iterative /ˈɪtərətiv/ adj. 迭代的

J

jailbreak /ˈdʒeɪlbreɪk/ n. 越狱 vt. (对手机，特别是 iPhone 等电子产品所做的) 破解
Java Runtime Environment Java 运行时环境

just-in-time adj. 及时的
JVM Java 虚拟机

K

kernel /ˈkɜːnl/ n. 内核
kit /kɪt/ n. 工具

knowledge representation 知识表达

L

label /ˈleɪbl/ v. 标志；贴标签

LAN 局域网

latency /ˈleɪtənsi/ n. 延迟;时延
laptop /ˈlæptɒp/ n. 笔记本电脑
latency /ˈleɪtnsi/ n. 延迟
ledger /ˈledʒə/ n. 账本
legacy /ˈleɡəsi/ n. 旧的;遗产
legal contracts 法律合同
leisure /ˈleʒə/ n. 空闲
Lenovo /lɪˈnəʊvəʊ/ n. 联想(公司名)
leverage /ˈliːvərɪdʒ/ v. 利用
lexical /ˈleksɪkl/ adj. 词汇的
library /ˈlaɪbrəri/ n. 库
license /ˈlaɪsns/ n. 许可证
lieu /ljuː/ n. 代替;场所
likeability /ˌlaɪkəˈbɪlətɪ/ n. 可爱
line-at-a-time 行式
linguistics /lɪŋˈɡwɪstɪks/ n. 语言学
livelihood /ˈlaɪvlihʊd/ n. 民生;生计
load balancing 负载平衡
localization /ˌləʊkəlaɪˈzeɪʃən/ n. 定位;地方化
locomotion /ˌləʊkəˈməʊʃn/ n. 运动
logician /ləˈdʒɪʃn/ n. 逻辑学家
logistics /ləˈdʒɪstɪks/ n. 物流

M

machine learning 机器学习
Mafia /ˈmæfiə/ n. 黑手党
mainframe /ˈmeɪnfreɪm/ n. 主机;大型机
maintenance /ˈmeɪntənəns/ n. 维护;维修
malfunction /ˌmælˈfʌŋkʃn/ n. 故障
malicious /məˈlɪʃəs/ adj. 恶意的
malware 恶意软件;流氓软件
MAN 城域网
metaverse /ˈmetəvɜːs/ n. 元宇宙;虚拟世界
manipulation /məˌnɪpjuˈleɪʃn/ n. 操作
maze /meɪz/ n. 错综复杂;迷惑
M-commerce 移动商务
mechanically /mɪˈkænɪkəli/ adv. 机械地
mechanism /ˈmekənɪzəm/ n. 机制
mediator /ˈmiːdɪeɪtə/ n. 中界者;调停者
megabyte /ˈmeɡəbaɪt/ n. 兆字节
mesh /meʃ/ n. 网状;网格
message /ˈmesɪdʒ/ n. 报文
metaphor /ˈmetəfə/ n. 隐喻;比喻
metered /ˈmiːtəd/ adj. 计量的
methodologies /ˌmeθəˈdɒlədʒi/ n. 方法论
metropolitan /ˌmetrəˈpɒlɪtən/ n. 大都市的
middleware /ˈmɪdlweə/ n. 中间件
migration /maɪˈɡreɪʃn/ n. 迁移
mileage /ˈmaɪlɪdʒ/ n. 里程
MIMD 多指令多数据流
miniaturization /ˌmɪnɪtʃəraɪˈzeɪʃn/ n. 微小型化
mission-critical 任务关键型的
mobile /ˈməʊbaɪl/ adj. 移动的 n. 手机
modular /ˈmɒdjʊlə/ adj. 模块化的
Mosaic /məʊˈzeɪɪk/ n. 浏览器名
motherboard /ˈmʌðəbɔːd/ n. 主板
multi-core 多核
multifaceted /ˌmʌltɪˈfæsɪtɪd/ adj. 多方面的
multitasking /ˈmʌltɪtɑːskɪŋ/ 多任务处理
mundane /mʌnˈdeɪn/ n. 平凡的事
Muslim /ˈmʊzlɪm/ adj. 穆斯林的

N

nanosecond /ˈnænəʊˌsekənd/ n. 纳秒；十亿分之一秒
nanotechnology /ˌnænəʊtekˈnɒlədʒi/ n. 纳米技术
nationwide /ˌneɪʃnˈwaɪd/ adj. 全国性的
NAS 网络连接存储
Navigator /ˈnævɪgeɪtə/ n. 浏览器名（领航员）
negligible /ˈneglɪdʒəbl/ adj. 微不足道的
negotiate /nɪˈgəʊʃieɪt/ v. 协商
Netscape /ˈnetˌskeɪp/ n. 美国网景公司
network /ˈnetwɜːk/ n. 网络
Network Access Layer 网络接口层
neural network 神经网络
neurobiology /ˌnjʊərəʊbaɪˈɒlədʒi/ n. 神经生物学
neuroscience /ˈnjʊərəʊsaɪəns/ n. 神经科学
NIC 网络接口卡
nonetheless /ˌnʌnðəˈles/ adv. 虽然如此
nonvolatile /ˌnɒnˈvɒlətaɪl/ adj. 非易失性的
numbering /ˈnʌmbərɪŋ/ n. 编号方式

O

object-based storage 基于对象的存储
object-oriented 面向对象的
obligation /ˌɒblɪˈgeɪʃn/ n. 合约；责任
open network 开放网络
open-source development 开源软件的开发
open-source 开放源码的；开源
operator /ˈɒpəreɪtə/ n. 操作员
organically /ɔːˈgænɪkli/ adv. 有机地
OS 操作系统
outage /ˈaʊtɪdʒ/ n. 断电；停机
outlier analysis 孤立点分析
outwit /ˌaʊtˈwɪt/ vt. 以智取胜；以计击败
overlap /ˌəʊvəˈlæp/ v. 重叠；相交
overload /ˌəʊvəˈləʊd/ vt. 使超负荷
oversight /ˈəʊvəsaɪt/ n. 监管；负责
overwhelmingly /ˌəʊvəˈhwelmɪŋli/ adv. 无法抵抗地

P

PaaS 平台即服务
packet switched 分组交换
packet /ˈpækɪt/ n. 分组
paging /ˈpeɪdʒɪŋ/ n. 分页
PAN 个域网
panoramic /ˌpænəˈræmɪk/ adj. 全景的
parallel /ˈpærəlel/ adj. 并行的
parity /ˈpærəti/ n. 奇偶
partitioning /pɑːˈtɪʃnɪŋ/ n. 分块
patent /ˈpæt(ə)nt/ n. 专利
pattern /ˈpætn/ n. 模式
pattern recognition 模式识别
pay-as-you-go 按使用付费
peak /piːk/ n. 峰值
peak performance 峰值性能
peer-to-peer 对等网；点对点

perception /pəˈsepʃn/ n. 感知
peripheral /pəˈrɪfərəl/ n. 外部设备
persistent /pəˈsɪstənt/ adj. 持久的
personalization /ˌpɜːsənəlaɪˈzeɪʃn/ n. 个性化
petaflops /ˈpiːtəflɒps/ n. 千万亿次；每秒千万亿次浮点运算
Pfize 美国制药公司
pharmaceutical /ˌfɑːməˈsjuːtɪkl/ n. 药物
philosophy /fɪˈlɒsəfi/ n. 哲学
phishing /ˈfɪʃɪŋ/ n. 网络仿冒；网络钓鱼
piconet /pɪkˈwʌnet/ n. 微微网
pipe /paɪp/ n. 管子；管道
pipelining 流水线
pirated versions 盗版
pivotal /ˈpɪvətl/ adj. 关键的
plaintext /pleɪnˈtekst/ n. 明码文本
plane display 平面显示
platform /ˈplætfɔːm/ n. 平台
port /pɔːt/ n. 端口
portable /ˈpɔːtəbl/ adj. 轻便的；手提的
power socket 电源插座
preference /ˈprefrəns/ n. 偏爱；优先权
preliminary /prɪˈlɪmɪnəri/ adj. 初步的
private cloud 私有云

proactive /prəʊˈæktɪv/ adj. 积极的；前瞻性的
probabilistic /ˌprɒbəbəˈlɪstɪk/ adj. 概率的
processor /ˈprəʊsesə/ n. 处理器
profile /ˈprəʊfaɪl/ n. 介绍；外形
prohibitively /prəʊˈhɪbɪtɪvli/ adv. 过高地
proliferation /prəˌlɪfəˈreɪʃn/ n. 增殖；扩散
prompt /prɒmpt/ n. 提示符
pronounced /prəˈnaʊnst/ adj. 明显的
propositional /ˌprɒpəˈzɪʃənl/ adj. 命题的
proprietary /prəˈpraɪətri/ adj. 专有的
prototype /ˈprtəʊtətaɪp/ n. 原型
prototyping /ˈprəʊtəʊtaɪpɪŋ/ n. 原型设计制作
provision /prəˈvɪʒn/ v. 调配；供应
pseudonymous /suːˈdɒnɪməs/ adj. 匿名的
psychology /saɪˈkɒlədʒi/ n. 心理学
public cloud 公共云
public-key encryption 公钥加密
punctuation /ˌpʌŋktʃʊˈeɪʃn/ n. 标点

Q

quantum /ˈkwɒntəm/ n. 量子
quantum effect 量子效应
qubit /ˈkjubɪt/ abbr. (quantum bit) 量子位
query /ˈkwɪəri/ v. 查询

questionnaire /ˌkwestʃəˈneə/ n. 调查表；调查问卷
quo /kwəʊ/ n. 现状
quote /kwəʊt/ n. 报价

R

rack /ræk/ n. 机架；货架
RAM 随机存储器
rational unified process 统一软件过程

realm /relm/ n. 领域；范围
real-time 实时
reasoning /ˈriːznɪŋ/ n. 推理

Reasoning Under Uncertainty 不确定性推理
redundancy /rɪˈdʌndənsi/ n. 冗余
refine /rɪˈfaɪn/ vt. 提炼；精炼
reflective /rɪˈflektɪv/ adj. 反射的
register /ˈredʒɪstə/ n. 寄存器
registration /ˌredʒɪˈstreɪʃn/ n. 注册
regulation /ˌreɡjʊˈleɪʃn/ n. 法规
relinquish /rɪˈlɪŋkwɪʃ/ vt. 放弃
relish /ˈrelɪʃ/ v. 品尝；欣赏
reload /ˌriːˈləʊd/ n. 重新加载
rendering /ˈrendərɪŋ/ n. 渲染
repeater /rɪˈpiːtə(r)/ n. 中继器
replicate /ˈreplɪkeɪt/ vt. 复制
repository /rɪˈpɒzɪtri/ n. 仓库
resembling /rɪˈzemblɪŋ/ v. 类似；像
reservoir /ˈrezəvwɑː/ n. 储藏
reside /rɪˈzaɪd/ vi. 属于；留驻
resilience /rɪˈzɪliəns/ n. 弹性
resistance /rɪˈzɪstəns/ n. 抵抗；反对
responsiveness /rɪˈspɒnsɪvnɪs/ n. 敏感性
retail /ˈriːteɪl/ n. 零售业
retrieve /rɪˈtriːv/ v. 检索
RFID 射频识别
rigged /rɪɡd/ adj. 装配的
rigorous /ˈrɪɡərəs/ adj. 严厉的；严密的
rigorous /ˈrɪɡərəs/ adj. 严密的
Robotics /rəʊˈbɒtɪks/ n. 机器人学
robust /rəʊˈbʌst/ adj. 强健；鲁棒
rogue /rəʊɡ/ n. 流氓；无赖
ROI /rwɑː/ n. 投资回报率
router /ˈruːtə/ n. 路由器

S

SaaS 软件即服务
SAN 存储区域网络
scalable /ˈskeɪləbl/ adj. 可扩展的
scheduler /ˈʃedjuːlə/ n. 调度程序
scheduling /ˈʃedjuːlɪŋ/ n. 调度
scripted testing 脚本测试
SD memory card 安全数字存储卡
seamlessly /ˈsiːmlɪsli/ adv. 无缝的
semantic /sɪˈmæntɪk/ adj. 语义的
sensor /ˈsensə/ n. 传感器
sequential pattern analysis 序列模式分析
serial /ˈsɪəriəl/ n. 串口；串行
sharper /ˈʃɑːpə(r)/ adj. 清晰的
shrink /ʃrɪŋk/ v. 减少
Siemens /ˈsiːməns/ 西门子公司
sift /sɪft/ v. 审查；筛查
Silicon Valley 硅谷
SIMD 单指令多数据
simulate /ˈsɪmjuleɪt/ v. 仿真；计算机模拟
simultaneously /ˌsɪməlˈteɪniəsli/ adv. 同时地
SISD 单指令单数据流
SLA 服务级别协议
slip into 塞进
slogan /ˈsləʊɡən/ n. 标语；口号
slot /slɒt/ n. 槽
smartphone /ˈsmɑːtfəʊn/ n. 智能手机
SMS 手机短信服务
Solaris /ˈsəʊlərais:/ 操作系统名
solid-state memory 固态存储器
space /speɪs/ n. 空白
spam /spæm/ n. 垃圾邮件
span /spæn/ vt. 跨越
spectral /ˈspektrəl/ adj. 光谱的
spectrum /ˈspektrəm/ n. 光谱；频谱
Sputnik /ˈspʌtnɪk/ n. (苏联)人造地球卫星

square /skweə/ n. 正方形；平方
stack /stæk/ n. 栈；堆栈
statement /ˈsteɪtmənt/ n. 语句
state identification 国家标识
statistics /stəˈtɪstɪks/ n. 统计学
stimuli /ˈstɪmjʊlaɪ/ n. 刺激因素
storage array 存储阵列
storage /ˈstɔːrɪdʒ/ n. 存储
stored-program concept 存储程序
stripped-down 简装的
subroutine /ˈsʌbruːtiːn/ n. 子程序
subscription /səbˈskrɪpʃn/ n. 预定
subtle /ˈsʌtl/ adj. 微妙的

Sugon /ˈsugən/ n. 中科曙光
suite /swiːt/ n. 组件
superimpose /ˌsuːpərɪmˈpəʊz/ vt. 叠加
suppression /səˈpreʃn/ n. 抑制；压制
sustained /səˈsteɪnd/ adj. 持久的
swarm /swɔːm/ n. 聚合；群
swiftly /ˈswɪftli/ adv. 敏捷地；快速地
switch /swɪtʃ/ n. 交换机
symposium /sɪmˈpəʊziəm/ n. 研讨会
synchronise /ˈsɪŋkrənaɪz/ v. 同步；同时发生
synthesis /ˈsɪnθəsɪs/ n. 综合

T

tablet /ˈtæblɪt/ n. 平板电脑
tactic /ˈtæktɪk/ n. 战术
tactile /ˈtæktaɪl/ adj. 触觉的
tadpole /ˈtædpəʊl/ n. 蝌蚪
tag /tæg/ n. 标签
tailor /ˈteɪlə/ vt. 特制
tamper /ˈtæmpə/ v. 篡改
tape reel 磁带盘
target /ˈtɑːgɪt/ n. 目标；目的
tenant /ˈtenənt/ n. 租户
Tennessee /ˌtenəˈsiː/ n.（美国）田纳西州
teraflops /ˈterəflɒps/ n. 万亿次；每秒万亿次浮点运算
term /tɜːm/ vt. 把……称为
terrain /ˈtereɪn/ n. 地形
territory /ˈterətri/ n. 领土；管辖区
test cases 测试案例
timing /ˈtaɪmɪŋ/ n. 定时；同步
the application layer 应用层
the State Council 国务院
the internet layer 网际层

the link layer 链路层
the transport layer 运输层
theoretical /ˌθɪəˈretɪkl/ adj. 理论的
thin client 精简客户端
throughput /ˈθruːpʊt/ n. 吞吐量
timeout /taɪmˈaʊt/ n. 超时
timesharing /ˈtaɪmʃeərɪŋ/ 分时（操作）
token /ˈtəʊkən/ n. 令牌；代币
tolerance /ˈtɒlərəns/ n. 容忍
toolkit /ˈtuːlkɪt/ n. 工具包
topology /təˈpɒlədʒɪ/ n. 拓扑学
traceability /ˌtreɪsəˈbɪləti/ n. 可追溯性
traffic /ˈtræfɪk/ n. 流量；交通
training example 训练样本
transistor /trænˈsɪstə/ n. 晶体管
translator /trænsˈleɪtə/ n. 翻译器；翻译程序
transparency /trænsˈpærənsɪ/ n. 透明性
transparent /trænsˈpærənt/ adj. 透明的
trial and error 反复试验
trial-and-error 不断摸索
trust /trʌst/ n. 信任

trustworthiness /ˈtrʌstwɜːðinəs/ n. 可信度

turbo /ˈtɜːbəʊ/ n. 涡轮

U

ubiquitous /juːˈbɪkwɪtəs/ adj. 无所不在的
ubiquity /juːˈbɪkwəti/ n. 到处存在；普遍性
umbrella /ʌmˈbrelə/ adj. 总称的；总括的
unauthorized /ʌnˈɔːθəraɪzd/ adj. 未经制授权的
unavailability /ˌʌnəˌveɪləˈbɪlɪti/ n. 无效
undertaking /ˌʌndəˈteɪkɪŋ/ n. 任务；事业
Unilever 联合利华公司
Unimate /ˈjuːnɪmeɪt/ 尤曼特；通用机器手
uninterrupted /ˌʌnɪntəˈrʌptɪd/ adj. 不间断的
unison /ˈjuːnɪzn/ n. 一致；协调

unobtrusive /ˌʌnəbˈtruːsɪv/ adj. 不突出的
unstructured data 非结构化数据
unveiled /ʌnˈveɪld/ adj. 公布于众的
upend /ʌpˈend/ v. 颠覆
up-front 前期
uphill /ʌpˈhɪl/ adj. 向上的
upload /ʌpˈləʊd/ vt. 上传
usability testing 可用性测试
usability /ˌjuːzəˈbɪlɪti/ n. 可用性；适用性
utility program 实用程序
utility /juːˈtɪləti/ n. 功用；效用；工具

V

vacuum /ˈvækjuəm/ n. 真空
value /ˈvæljuː/ n. 价值
vandalism /ˈvændəlɪzəm/ n. 故意破坏；捣毁
variable /ˈveərɪəbl/ n. 变量
variation /ˌveərɪˈeɪʃn/ n. 变化
vehicle /ˈvɪəkl/ n. 汽车
vendor /ˈvendə(r)/ n. 供应商；卖主
ventilation /ˌventɪˈleɪʃn/ n. 通风设备
verification /ˌverɪfɪˈkeɪʃn/ n. 证明；证实
versatile /ˈvɜːsətaɪl/ adj. 多用途的
vibration /vaɪˈbreɪʃn/ n. 震动
victim /ˈvɪktɪm/ n. 受害者；受骗者
violate /ˈvaɪəleɪt/ vt. 违反
violet /ˈvaɪələt/ n. 紫色
virtual machine 虚拟机

virtual /ˈvɜːtʃʊəl/ adj. 虚拟的
virtualization /ˌvɜːtʃʊəlaɪˈzeɪʃn/ n. 虚拟技术
VLAN 虚拟局域网
ultra /ˈʌltrə/ adj. 超级；极端
VM 虚拟机
volatility /ˌvɒləˈtɪləti/ n. 易失性；挥发性
volume /ˈvɒljuːm/ n. 卷；量；体积
voluntary /ˈvɒləntri/ adj. 自愿的
volunteer /ˌvɒlənˈtɪə(r)/ n. 自(志)愿者
VSAN 虚拟存储区域网络
vulnerability /ˌvʌlnərəˈbɪləti/ n. 漏洞；弱点
vulnerable /ˈvʌlnərəbl/ adj. 易受攻击的；易受伤的

W

wafer /ˈweɪfə/ n. 薄片
Walmart /ˈwɔlmaːt/ n. 沃尔玛
WAN 广域网
warranty /ˈwɒrənti/ n. 担保
waterfall model 瀑布模型
weapon /ˈwepən/ n. 武器
well-liked 深受喜爱的
wizard /ˈwɪzəd/ n. 向导
workload /ˈwɜːkˌləʊd/ n. 工作负载
World Wide Web 万维网；环球网
wrestle /ˈresl/ v. 斗争

缩略词表

ACLs　Access Control Lists　访问控制列表
AGI　Artificial General Intelligence　通用人工智能
AI　Artificial Intelligence　人工智能
AIG　American International Group　美国国际集团
AOL　American Online　美国在线
AP　Access Point　访问接入点
API　Application Program Interface　应用程序编程接口
AR　Augmented Reality　增强现实
BASIC　Beginners All-Purpose Symbolic　BASIC 语言
　　　　Instruction Code
BCI　Brain-ComputerInterface　脑机接口
BD　Blu-Ray Disk　蓝光光盘
BOINC　Berkeley's Open Infrastructure for　伯克利开放式网络计算平台
　　　　Network Computing
CART　Classification And Regression Tree　分类与回归树
CN　Core Network　无线接入网络
C/S　Client/Server　客户/服务器
CASE　Computer Aided Software Engineering　计算机辅助软件工程
CD　Compact Disk　唱片，光盘
CLI　Command-Line Interface　命令行界面
COVID-19　Corona Virus Disease 2019　新型冠状病毒肺炎
CPU　Central Processing Unit　中央处理器
CRM　Customer Relationship Management　客户关系管理
CT　Computed Tomography　电子计算机断层扫描
DAI　Distributed Artificial Intelligence　分布式人工智能
DBMS　Database Management System　数据库管理系统
DL　Deep Learning　深度学习
DM　Data Mining　数据挖掘
DNA　Deoxyribonucleic Acid　脱氧核糖核酸
DOS　Disk Operating System　磁盘操作系统
DoS　Denial of Service　拒绝服务
DVD　Digital Versatile Disk　数字化视频光盘
EC　Electronic Commerce　电子商务

ENIAC   Electronic Numerical Integrator and   电子数字积分计算机，埃里阿克计算机
         Calculator
EU   European Union   欧盟
FAT   file allocation table   文件分配表
FC   Fibre Channel   光纤通道
FCoE   Fibre Channel over Ethernet   基于以太网的光纤通信技术网络
FOSS   Free and Open Source Software   免费的开源软件
FTP   File Transfer Protocol   文件传输协议
FTPS   FTP-over-SSL   相当于加密版的 FTP
GB   GigaByte   千兆字节
GIF   Graphics Interchange Format   一种图片格式
GPL   General Public License   通用公共许可证
GPRS   General Packet Radio Service   通用分组无线业务
GPS   Global Positioning Systems   全球定位系统
GSM   Global Systems for Mobile   全球移动通信系统
HDC   Huawei Developer Conference   华为开发者大会
HMS   Huawei Mobile Services   华为移动服务
HTML   Hyper Text Markup language   超文本标记语言
HTTP   Hypertext Transfer Protocol   超文本传输协议
HTTPS   Hyper Text Transfer Protocol over Secure   安全的 HTTP
         Socket Layer
I/O   Input/Output   输入输出设备
IaaS   Infrastructure as a Service   基础架构即服务
IBM   International Business Machines Corp.   （美国）国际商用机器公司
ID3   Iterative Dichotomizer 3   一种构造决策树的算法
IDE   Integrated Development Environment   集成开发环境
IE   Internet Explorer   微软公司开发的浏览器
IEC   International Electro Technical Commission   国际电工委员会
IEEE   Institute of Electrical and Electronics Engineers   电气和电子工程师协会
IETF   Internet Engineering Task Force   互联网工程任务小组
IM   Instant Messaging   即时消息
IOS   Internet Operating System   互联网操作系统
IoT   Internet of Things   物联网
IP   Internet Protocol   网际协议
ISC   International Supercomputer Conference   国际超级计算机大会
ISO   International Organization for Standardization   国际标准化组织
IT   Information Technology   信息技术
J2ME   Java 2 Micro Edition   用于嵌入式产品的 Java 开发平台

J2SE　Java 2 Standard Edition　Java2 标准版
Java EE　Java Enterprise Edition　Java 企业版
JDK　Java Development Kit　Java 开发工具包
JIT　Just-In-Time　及时的
JPEG　Joint Photographic Experts Group　一种图片格式
JRE　Java Runtime Environment　Java 运行时环境
JTC　Joint Telecommunications Committee　＜美国＞联合电信委员会
JVM　Java virtual machine　Java 虚拟机
KDD　Knowledge Discovery in Database　知识发现
LAN　Local Area Network　局域网
LISP　List Processing　一种人工智能语言
MAC　Mandatory Access Control　介质访问控制层
MAN　Metropolitan Area Network　城域网
MAS　Multi-Agent System　多智能体系统
MID　Mobile Internet Devices　移动互联网设备
MIMD　Multiple Instruction Multiple Stream　多指令多数据流
MOOC　Massive Open Online Courses　慕课
MS　Microsoft　微软
MSN　Messenger　即时通信软件
NAS　Network Attached Storage　网络连接存储
NASA　National Aeronautics and Space Administration　美国航空航天局
NCSA　National Center for Supercomputing Applications　美国国家超级计算应用中心
NFV　Network Functions Virtualisation　网络功能虚拟化
NIC　Network Interface Card　网络接口卡
NIST　National Institute of Standards and Technology　美国国家标准与技术研究院
NPC　National People's Congress　全国人民代表大会
ONF　Open Networking Foundation　开放网络基金会
OOP　Object Oriented Programming　面向对象的程序设计
OS　Operating System　操作系统
P2P　Peer-to-Peer　点对点
PaaS　Platform-as-a-Service　平台即服务
PAN　Personal Area Network　个域网
PC　Personal Computer　个人计算机
PDA　Personal Digital Assistant　个人数字助理
PROLOG　Programming in Logic　面向演绎推理的逻辑型程序设计语言
PSP　Play Station Portable　索尼公司开发的新型多功能掌机
RAM　Random Access Memory　随机存储器
RAN　Radio Access Network　无线接入网络

RFC   Request for Comments   请求评议
RFID   Radio Frequency Identification   射频识别技术
RL   Reinforcement Learning   增强学习
RUP   Rational Unified Process   统一软件过程
SaaS   Software-as-a-Service   软件即服务
SAN   Storage Area Network   存储区域网络
SD card   Secure Digital card   安全数码卡
SDHC   Secure Digital High Capacity   高容量安全数字卡
SDK   Software Development Kit   软件开发工具包
SDXC   Secure Digital Extended Capacity   扩展容量安全数字卡
SIM   Subscriber Identity Module   用户身份识别卡
SIMD   Single Instruction Multiple Data   单指令多数据
SISD   Single Instruction Single Data   单指令单数据流
SLA   Service Level Agreement   服务级别协议
SMS   Short Message Service   手机短信服务
SMSC   Short Message Service Center   短消息服务中心
SOA   Service Oriented Architecture   面向服务体系
SSH   Secure Shell   安全壳协议
SSL   Security Socket Layer   加密套接字协议层
SVM   Support Vector Machines   支持向量机
TCP   Transfer Control Protocol   传输控制协议
UAV   Unmanned Aerial Vehicle   无人战斗机
UMTS   Universal Mobile Telecommunications System   通用移动通信业务
URL   Uniform Resource Locator   统一资源定位器
VDC   Virtualized Data Center   虚拟数据中心
VLAN   Virtual Local Area Network   虚拟局域网
VM   Virtual Machine   虚拟机
VR   Virtual Reality   虚拟现实
VSAN   Virtual Storage Area Network   虚拟存储区域网络
WAN   Wide Area Network   广域网
WAP   Wireless Application Protocol   无线应用协议
WIC   WAN Interface Card   广域网接口卡
WML   Wireless Markup Language   无线标记语言
WORA   Write Once, Run Anywhere   一次编写,到处运行
WTLS   Wireless Transport Layer Security   无线传输层安全
WWW   World Wide Web   万维网;环球网
XP   Extreme Programming   极限编程

参考文献

[1] Brookshear J G. Computer Science An Overview[M]. 11th Ed. 北京：人民邮电出版社,2013.

[2] Stallings W. Operating System—Internals and Design Principles[M]. 6th Ed. 北京：电子工业出版社,2010.

[3] Somasundaram G, Shrivastava A. Information Storage and Management[M]. 2nd Ed. 北京：人民邮电出版社,2013.

[4] Joseph S D, Redish C J. A Practical Guide to Usability Testing[M]. London：Intellect Books, 1999.

[5] Nilsson J N. Artificial Intelligence：A New Synthesis[M]. 北京：机械工业出版社,1998.

[6] Ertel W. Introduction to Artificial Intelligence[M]. 2nd Ed. Springer. 2018.

[7] 钟静,吴鸿娟,郭皎. 计算机英语[M]. 北京：清华大学出版社,2015.

[8] 刘艺,王春生. 计算机英语[M]. 4版. 北京：机械工业出版社,2013.

[9] 王春生,刘艺,杨伟荣. 新编计算机英语[M]. 北京：机械工业出版社,2007.

[10] 宋德富,司爱侠. 计算机专业英语教程[M]. 3版. 北京：高等教育出版社,2008.

[11] ED informatics.History and Development of Robots[EB/OL]. London：Royal Veterinary College [2014-10-08]. http://www.edinformatics.com/math_science/robotics/robot1.htm.

[12] Merriam Webster. Internet of Things[EB/OL].[2022-02-15]. https://www.merriam-webster.com/dictionary/Internet of Things.

[13] Alex Chris.Top 10 Search Engines in the World[EB/OL].[2022-4-20]. https://www.reliablesoft.net/top-10-search-engines-in-the-world/.

[14] Bryan Lynn. What is the Metaverse Tech Companies Aim to Build？[EB/OL].[2022-02-15]. https://www.unsv.com/voanews/specialenglish/scripts/2021/09/13/0605/.

[15] Metaverse. Metaverse Ecosystem[EB/OL].[2022-02-10].https://mvs.org/.